History *of* Science Selections *from* ISIS

GENERAL EDITOR, Robert P. Multhauf

Early American Science

Early American Science

Brooke Hindle

Editor

Science History Publications
New York · 1976

First published in the United States by
Science History Publications
a division of
Neale Watson Academic Publications, Inc.
156 Fifth Avenue, New York 10010

First Edition 1976
Designed and manufactured in the U.S.A.

(CIP data follows page 213)

Publisher's Note: We should like to express our appreciation to ISIS for granting permission to reprint the articles in this volume. If the publishers have unwittingly infringed the copyright in any illustration reproduced they will gladly pay an appropriate fee on being satisfied as to the owner's title.

Contents

SCIENCE IN SOCIETY

Introduction

Early American Science has a surprisingly large representation in the many volumes of *Isis*, surprising because American science was not of major interest to the founder of the journal or to most of his followers. Moreover, even today, early American science is a lightly cultivated field, and with so little being published anywhere, the value of this collection is the greater. These selected articles have, in addition, a very special character; their historiography is important.

Isis was founded by George Sarton, a scientist who applied his enormous dedication and energy to the study of the history of science. He and a small band of scholars focused upon the early period of history, as a foundation for continuing studies. Sarton's own writings were of two sorts, the first, detailed, factual reports and bibliographical listings such as those contained in his monumental *Introduction to the History of Science*, and the second, exhortatory essays on such subjects as "the new humanism." He encouraged a positivistic and intellectual history of science or history of scientific ideas.

In its early years, *Isis* resembled more a scientific than a historical journal; its contents included many brief, technical reports upon small inquiries over a broad range of ancient, medieval, and early modern history. The key research reports and information were expected to appear in this medium, whereas most historical journals play a different role. On the whole, historians view books as the primary form of presentation for mature historical syntheses; historical journals offer the opportunity for circulating concepts in the process of development and for presenting finished essays which are of limited scope or which represent by-products of larger studies published separately. *Isis* became a historical journal and roundeu historical essays became more general.

Sarton's own students acquired much of his conspicuous dedication, but developed as well a broad sophistication and sensitivity to the whole terrain of history. With new standards of rigor, they have flourished, but only a few have ventured into American history. Even so, their influence, and that of Sarton himself, pervades this collection and gives it a special flavor.

Still more, however, this collection reflects another viewpoint which has grown with the years. Many historians have given lip service to the assertion that history, specifically American history, must include science and technology as an integral part of the general synthesis. Yet, only recently have specific advances been made toward this end.

Arthur M. Schlesinger, long a colleague of Sarton, encouraged the history of science with a very different objective. A pioneering social historian, he always understood science and technology to be necessary elements in any satisfactory understanding of American history. Although his own writings on the history of science were included within larger syntheses, he inspired students who gave their primary attention to science.

Schlesinger and his students always had in mind the large view of history. They never doubted that competent history required the best, detailed knowledge of any area of science brought under serious scrutiny. They did not believe that their work was "externalist" or concerned with the social dimensions of science to the neglect of its "internal" history. Some, indeed, gave much attention to the specific character of scientific ideas, but without relinquishing concern for the many linkages of science.

Richard H. Shryock had a still different influence upon the study of the history of science. With a traditional historical training, he entered the fields of "American and medical history," insisting that his study of American history was not limited to medicine and his writings in medical history were not bounded by the United States. In fact, his research upon medicine and public health was only partially within or related to the history of science. As Schlesinger was an exponent, Shryock was a practitioner of comparative history, and he succeeded in carrying medical history beyond the biographical, intellectual, and internal to the most comprehensive aspirations of understanding. While worrying with the most pesky details, he never ceased to ask the big questions.

The actual route by which an expanding interest in American science grew in *Isis* is curious. A large majority of the American articles is biographical—why? Biography, of course, is an intimate part of and partner in historical writing. All historians, but especially those who think of themselves first as humanists, must be concerned with biography and with the human story. It is equally notable that scientists, physicians, engineers, and others who turn to the history of their own professional fields tend to make use of the biographical approach. They are usually anxious to emphasize individual contributions and can identify with historical figures in fields akin to their own. Biography too, whether deceptively or not, seems to offer one of the most direct approaches to the history of science, an approach which automatically limits the focus of study and permits attention to both the internal history of ideas and their social setting and meaning.

More specifically, biography may have provided a needed validation of the acceptability of an American subject in a community which understood the higher priority of working on science in its seminal periods and in the places where the critical advances were made. This certainly had something to do with the heavy concentration upon Benjamin Franklin. The question had long been asked of those who expressed interest in American science, "Where is the American Newton?" "Where is the American Darwin?" Against an extraordinary dearth of recognized models of American achievement in science, Franklin had long been the conspicuous example of one American whose work in science might possibly be worthy of attention.

This scale of values was cut in stone in 1916 in the new buildings of the Massachusetts Institute of Technology on the Cambridge side of the Charles. The monumental architecture offered opportunity to inscribe in gigantic letters the names of the ten greatest scientists of all time. Beside Aristotle, Galileo, and Newton, was there really an American who deserved mention? The choice of Benjamin Franklin at that point of time was unavoidable, and it was accepted by both scientists and historians.

From the earliest days of *Isis*, therefore, Franklin was a possible subject. More notes, short articles, and essays on Franklin have appeared in the journal than on any other American, most of them by scholars who encountered him while pursuing other related studies. Conway Zirkle, for example, was a botanist who actively studied the history of his profession, J.A. Leo Lemay and Alfred Owen Aldridge are literary scholars who intersected Franklin's science, and Bernard S. Finn is a historian of science, studying the history of electricity.

The only true Franklin scholar represented is I. Bernard Cohen, who was one of Sarton's earliest students. Of his notes published here, one appeared just after his critical edition of *Benjamin Franklin's Experiments* and the other two just before his *Franklin and Newton*. His contributions, like several of the others, establish Franklin's relationship to other major figures and to concepts which developed over a period of time.

The biographical articles in this collection introduce secondary figures in American science. Each of them had not only a local importance but an international impact and recognition, even though science commanded no more than a minor portion of their effort. All of the authors published these biographical summaries as by-products of larger efforts. Frederick G. Kilgour was then engaged in studying science in early New England and Everett Mendelsohn was pursuing graduate studies in the history of science, in which early American science was a minor refrain. Both Frederick B. Tolles and Leonard Tucker published book-length biographies of their subjects in which they did not emphasize science primarily. Yet, each became convinced that the scientific contribution of his subject was important, and each prepared his best analysis of that work for *Isis*.

Of course, to group a few studies under "Scientific Ideas," is deceptive since all of the papers in this book are in some measure concerned with scientific ideas. Mel Gorman's article, however, is one of the most important fruits of his study of the transmission of ideas and Robert Siegfried's represents an American episode within his broad concern with chemistry and concepts of matter. The essay I did with Helen M. Hindle is a full analysis of an experimental inquiry which was of limited importance in my book-length biography. Kentwood D. Wells, curiously, gives the best analysis yet published of the ideas of an early Charlestonian, an "internalist" study of a man first publicized by Richard H. Shryock who was often regarded as an "externalist."

The three final essays are concerned with science in society. Mine, typically, is institutional and reflects Shryock's influence in the effort to probe social motivation in science. John C. Greene's essays equally demonstrate the Schlesinger influence; both of them forecast elements of a forthcoming book on science in Jeffersonian America. All three essays ask questions about the larger meaning of science in early America.

Unlike John C. Greene's two papers, however, the present collection of essays as a whole does not represent a prelude or introduction to research and publication soon to come. It contains a full index to a completed, historiographical episode in the study of early American science. Strikingly, only Greene and Wells seem still to be working in this field of history, and neither of them remains wholly within it. Actually, most of the authors represented here were never primarily interested in early American science. Several who were have shifted their major effort to other periods, to non-American science, or to related fields such as technology. Only a few new figures have entered the field. Why? What has happened? What is the meaning of this collection?

The answers are remarkably well spelled out in this discrete, historiographical archive. Before the Second World War, nothing of consequence on early American science appeared in *Isis*; indeed, nothing much was written anywhere. Sarton's introductory essay accurately defines the prevailing lack of encouragement. With the end of the war, however, both Schlesinger and Shryock made significant efforts to stimulate the study of American science—including specifically early American science. They probably voiced a rising sense of need of which they were merely spokesmen.

Before the 1950s, *Isis* carried only scattered early American articles. Then, suddenly, something appeared every year beginning in 1954. By 1961, two-thirds of the papers reprinted here had been published; more than a quarter came out in the single year, 1955. Then, the flow almost totally stopped—only three articles in the last dozen years.

Clearly the war marked a turning point and brought new demands. Directors of graduate studies who were sensitive to the need, encouraged their students, some of whom began to produce results in the 1950s. Whether a product of the same movement or not, I. Bernard Cohen came out of the war, having just completed his doctoral study of Franklin, and in 1953 became editor of *Isis*. Still actively engaged in Franklin research, he published elements of his own work and readily received early American contributions from others. Because of the encouragements of the time and the receptivity of *Isis*, early American science flourished.

Why did it recede in the 1960s? Several related developments occurred. The richest fields in American science were found in later American history and younger students tended to turn their attention there. This trend was encouraged by events at the end of the decade which emphasized relevance and stressed the human dimensions and social meaning of science. Already, there was a tendency to follow the main stream of science, and that can seldom be done within American history except in the later period.

Precisely because the present study of the history of science does not place early American science at its center, this collection is important. It records a collective search for better understanding of the role and character of science during a seminal period of American history. Many different approaches were involved and new insights were developed which have continuing validity. The present essays are representative of the best results attained in an extended inquiry. The search continues, even though now in a low key. The need is permanent.

BROOKE HINDLE
Smithsonian Institution

Preface to Volume VIII

(Being the third volume published under the auspices of the History of Science Society)

At the moment of opening the eighth volume of *Isis,* I realize more deeply than ever how it still falls short of the ideal of its founder and patrons. In spite of the faithful collaboration of so many distinguished scholars of many countries, I feel that we have not yet redeemed half the promises which we have made. However we are undoubtedly on the road to improvement, for it can easily be shown, I believe, that each volume is superior to the previous one; there is thus good reason to be trustful. It should be remembered that the aim of our journal is twofold; first, to provide the students of the History of Science with an indispensable tool, and to help them construct their own scientific apparatus; second, to give to a larger audience as much information as possible on the new intellectual movement whereof our special studies form, so to say, the core, and to diffuse more accurate views on the progress of civilization.

*
* *

A history of civilization, focussed, as it ought to be, upon the development of science, is not the history of any single province of humanity; it is the history of the whole of mankind, the astounding account of man's progress from the abyss of bestiality towards a goal which is almost divine. The History of Science is essentially an international history; hence the necessity of an international cooperation to accomplish our purpose. Yet, however desirable the collaboration of many foreign scholars each of whom brings not simply his own treasure of information but also his special point of view, the polyglottism of our journal must be considered, not as an advantage, but as a real evil to be overcome as soon as feasible. It is the Editor's hope to be able some time to publish the whole of *Isis* in one or at most in two languages. We are now using five languages (English, French, German, Italian, Latin), and the circle of our readers is thus considerably restricted. There is no greater fallacy than to believe that a polyglot review is more international than one written in a single language; it is, in fact, far less international, for there are undoubtedly far more people in any country who know, say English, than there are people who know English, plus French, plus German, plus Italian, plus Latin. Each language reduces materially the number of readers. The internationalism of an organ is not determined by the language but by the subject. Some nations publish propaganda literature in many languages; that literature remains nationalistic no matter in how many languages it is made to appear.

It may be pointed out that the History of Science down to the XVIIIth century and even down to the middle of the XIXth century, is largely European and Oriental history, and thus that it is far easier for American, than for European, scholars to be unbiassed and to take a truly international view of the whole field. Moreover I spare no pains to have every point of view represented as far as this can be done in a scholarly fashion. It is of course impossible to give in any journal, even if it were much larger than *Isis,* the history of every branch of science as it developed in every country. I must be content to publish the valuable material that comes my way or to solicit the contributions of scholars who are known to have made a careful study of a special subject. However, if it is impossible to publish fresh documents bearing on every angle of our vast discipline, we can at least give briefer or longer accounts of everything of importance, which has appeared anywhere in the world, and this, as the readers of *Isis* know, we are trying hard to do. Great efforts are made to bring to their attention everything of importance published in the five main languages above-mentioned. With regard to books or papers appearing in other languages, it is out of the question to obtain a similar completeness, nor is it at all desirable. We have every right to assume that scholars publishing papers in one of the smaller languages did not really wish to address the whole Republic of Letters but only their own countrymen. We have even the right to believe, until further notice, that their publications are not really important but are simply meant to diffuse in their linguistic province results which were already available in a large part of the civilized world by means of the greater languages. Yet the most important contributions are reviewed in *Isis,* irrespective of the language; for example the first seven volumes contain analyses and discussions of studies published in the following languages (and possibly in others) : Amharic, Arabic, Chinese, Danish, Dutch, Gaelic, Greek, Hebrew, Japanese, Polish, Portuguese, Russian, Spanish, Swedish. This is not a bad record, and the more so, I wish to emphasize it, that I feel under no special obligation to dig out works which their authors have chosen to hide from international audiences by publishing them in a language which the vast majority of scholars can not be expected to know. If their works remain unknown it is undoubtedly their fault, much more than ours.

Some of our shortcomings, of which nobody can be more deeply aware than the Editor, are probably due to the fact that our studies represent a new discipline, which is located at the borders of many others : science, history, philosophy, philology, religion, and is thus necessarily the source of many misunderstandings. I make serious attempts to satisfy the needs of all scholars attracted to our studies, from whatever direction they may have come to them, but because of this, I must necessarily fail to satisfy completely the needs of any special group. I beg scholars to keep this always in mind and to be patient. None of them can possibly be served as well as I would wish, because there are so many others whom I must serve at the same time and whose needs are entirely different.

Taking this into account, it speaks well for the intellectual elite of our time that our young society has been so successful from the very first year. Think of it : more than five hundred men and women, scattered all over the world, joined our flag at the first call. And yet I can not but feel that this is only a very humble beginning. If we realize how many people are enrolled in a legion of societies whose historical horizon is restricted to one very special subject, say genealogy or numismatics, or to a very small area, one single country or province or even community, is it unreasonable to hope that by and by a very large number of them will become interested in one of the noblest phases of the history of mankind ? There is every reason to hope that as soon as we have rid ourselves of that cumbersome polyglottism and improved our journal in various other ways, — as soon as we will truly deserve it, — our membership will not be a matter of hundreds but of thousands. In the meanwhile the greater success of local societies is natural enough. We are interested in, and love, our family first; then the people of our town or parish, people whom we know full well and whose habits and ways of thinking are nearest to our own; then, our fellow citizens; and thus very slowly do we enlarge the sphere of our loyalty and devotion. The opposite order would be unsound and suspicious. We are unable to trust people who profess to love the whole world but fail to love their neighbors. The latest phase of this evolution, the international one, is hardly begun and thus it would be surprising if a society devoted to the study of the progress of civilisation would be very successful at once. But the sphere of our influence is bound to increase, that is, if we justify the confidence of our readers.

* * *

Our greatest shortcoming perhaps is our failure to publish more articles of a simple and general nature which the majority of our members could read with pleasure. It is, however, extremely difficult to obtain articles combining literary excellence with sound scholarship and a sufficiently broad view of a vast subject. The scientific soundness of such articles must be even greater than that of the more special papers which are read only by experienced scholars, for these can be trusted to read between the lines and take some statements, if needs be, with a grain of salt. *The more popular an article, the more authoritative it must be.*

We are entitled to assume that our readers are in earnest, that they do not require to be amused or cajoled or flattered, that all they ask for is the most reliable information, whether they immediately need it or not, the best approach to truth, — an approach which must needs be very humble and very quiet. *Isis* is of necessity a very austere journal. Its essential duty is to uphold as high an ideal of scholarship as possible. Nowhere is this more needed than in the New World, where the very eagerness of people to be informed without having acquired sufficient means of weighing the soundness of the information imparted to them causes the cheapening of every thought, a gradual deterioration of the spirit. The degradation of intellectual energy entailed by a universal and uncritical curiosity is so great that it may seem sometimes impossible to resist it. Reason the more for all those who care for truth and disinterested research, to band themselves and make no concessions.

Many of the papers published in *Isis* are very technical and thereby unreadable save for very few scholars. It is probable that the only person who reads the whole of *Isis* is the Editor and he does not always enjoy it. He does not enjoy it because even he himself is not and cannot be equally interested in every part of the subject. The reviews are in general more readable, though in this case too, the needs of scholarship must be considered in the first place. The best way of upholding scholarship is to give adequate praise to every scholarly work and to denounce with equal vigor any piece of humbug. One of the greatest tragedies of our intellectual life (especially in America) is that reviews which are most valuable from the commercial point of view (that is, those appearing in newspapers and popular magazines) are often written by incompetent men who are just as likely to applaud unsound books and condemn sound ones as to do the contrary. More and more pains will be taken to correct such injustice and to push into the limelight, as far as it is in our power, those works which are positive, not negative, lasting, not ephemeral, contributions to knowledge.

Many subjects dealt with in *Isis* must seem of little importance to those readers who are not sufficiently equipped to understand their connections and implications. I beg them to trust for such matters the judgment of the editors. Scientific progress is a very slow and arduous process. It may take years of drudgery to justify one generalization. For the same reason hundreds of pages of *Isis* will seem hopelessly dry to many readers. This is not one of our many shortcomings but an unavoidable necessity. Generalizations may be all right on Sundays; they are not in order on working days. We hope to be able to publish sometimes a « holiday » number, but we do not promise it. We hope to publish popular articles, but most will needs be unpopular. Some people think that they are inspired all the time, but so much inspiration is rather suspicious; it is more probable that they are simply intoxicated or infatuated. One hour of divine inspiration in a life-time of work, one inspired page in a whole volume of *Isis*, that is enough, if we are wise enough to make the most of it.

Isis is a very austere journal. In spite — and perhaps because — of which, we expect that all those who love scholarship and truth for its own sake, will help us with all their might and main.

S. S. Rotterdam (H. A. L.). June 11, 1925.

GEORGE SARTON.

AN AMERICAN HISTORIAN LOOKS AT SCIENCE AND TECHNOLOGY

By Arthur M. Schlesinger

"History, as generally written, is but an account of the wars and contentions by which dynasties have striven for the mastery of nations. It imparts little or no information in respect to the social conditions or material progress of the people themselves. . . . Inasmuch, however, as that the nature, the institutions, and the administration of the American nation are different from all others, so must its history be in an entirely different style. . . . If we have no Alexander, or Caesar, or Bonaparte, or Wellington, to shine on the stormy pages of our history, we have such names as Franklin, Whitney, Morse, and a host of others, to shed a more beneficent lustre on the story of our rise. The means by which a few poor colonists have come to excel all nations in the arts of peace, and to astonish the people of Europe with their achievements through the development of their inventive genius, are true subjects for a history of the United States."

This passage has an authentically modern ring, but actually it appeared in the preface of a collaborative work entitled *Eighty Years' Progress of the United States*, published in 1866. Though American historians were put thus early on notice, in the eighty years since they have nevertheless continued for the most part to celebrate wars and political contentions — the Caesars and Wellingtons rather than the Franklins and Morses. The additions to our knowledge of science and technology have come largely from other sources, two in particular, either from specialists in the various branches of science or from that breed of authors called popularizers. Those who wrote with expert understanding generally talked above the heads of lay readers, while the popularizers, seeking romance and drama in every new conquest of Nature, generally overplayed the facts or garbled the underlying principles. Both sets of writers, moreover, have tended to look at their theme with blinders, failing to correlate scientific achievement with the broader movements of history — as though the investigator or inventor lived and labored in a social vacuum.

Finally, it should be noted that the bulk of this literature deals with technology or applied science, very little of it with theoretical or pure science. To be sure, the Americans, a practical-minded people, have scored most heavily in the applied field, preferring what they term useful research to useless research. But neither this fact nor the inherent difficulty of elucidating the abstruse justifies the failure to portray adequately the developments in pure science. It hardly needs to be said that so-called useless research often turns out in the end to be the most useful kind, for it constitutes a seedbed of endless utilitarian applications. Furthermore, as the United States has matured as a nation, it has contributed ever more importantly to science and technology at the theoretical level. The four-volume history of American science, now being planned by the American Council of Learned Societies, will, if carried out, go far toward filling this gap.

Professional scholars have been slow to study the progress of science and technology, partly because of their traditional penchant for political history, and partly because, even with the rise of social and intellectual history, they have been daunted by the unfamiliar and formidable subject matter. At certain points, it is true, they have touched upon epoch-making mechanical inventions which had self-evident economic consequences, but these occasional references fall far short of depicting the pervasive and continuing role of science in all ranges of American life. For this more comprehensive task most historians have felt inadequate because their specialized training did not fit them for it.

How is this barrier to be surmounted for the future? How can a program of graduate study be designed to accomplish for scholars of the next generation what has been lacking in the preparation of our own? The difficulties are many. On the one hand, professors of history must surrender jealously guarded vested interests in order to make room for the new type of subject matter. On the other hand, their colleagues from the laboratory must co-operate by packaging their wares for a new type of customer. Since the historical student is not planning to become a specialist in any scientific field, his needs call for a different kind of instruction. As I see it, the emphasis should be placed on the methods and unifying elements of all science, the differentiating principles which have brought about increasing fragmentation and the historic discoveries that underlay the pivotal advances. To the extent that the study focuses on the United States, it will be necessary further to separate the American strands from the world web of science.

This is doubtless a counsel of perfection, and while we await Utopia, much cries to be done. The historian, even when insufficiently informed as to the data of science and technology, can often perceive social implications and interrelations which specialists in those branches are unaware of or disregard. Here, in fact, lies the peculiar function of the historian: not so much to write the internal history of science as to trace the external connections of science and society. This relationship works both ways. Sometimes society motivates scientific discovery. Sometimes a new advance of science motivates social change.

Instances abound of the creative role of social conditions. As the old saying has it, necessity is the mother of invention. Benjamin Franklin in devising the famous fireplace stove explained that his attention had been directed to the need of a fuel-conserving burner by the growing scarcity and expense of wood as the forests were cut down. The invention of the cotton gin represented a different sort of necessity. This appliance might have been contrived much earlier than it was, for it involved very simple mechanical principles; but, as every student knows, it was not till the 1790's that the demands of the English textile industry revealed what a gold mine lay in short-staple cotton if it could be properly cleansed for the overseas market. A Connecticut schoolmaster sojourning on a Georgia plantation supplied the answer. A generation later, when Westerners were turning from subsistence tillage to commercial agriculture, the difficulty of securing hired hands called into being all the basic modern labor-saving farm implements: the steel plow, the reaper, the thresher and the rest. As American life grew more urbanized and complex, social incentives operated to yet different ends. It would be pointless to multiply cases, but suffice it to say that the great growth of cities after the Civil War brought forth in quick succession the elevated railroad, the cable car and the trolley car as solutions for the problem of traffic congestion, and also begot America's unique architectural innovation, the skyscraper. The influence of war in stimulating science and technology is so fresh in our minds as to require no comment except perhaps to remark that every major American war has had this effect and, further, that the results have appeared not only in death-dealing weapons and explosives, but also in notable gains in sanitation, medicine and surgery.

While these and countless other instances support the adage that necessity is the mother of invention, the historian is obliged to demur that the expected pregnancy doesn't always ensue. In pure science, where intellectual creativeness is the analogue of technological invention, the investigator tends to pursue the bent of his curiosity without reference to time and place. It would be difficult, for example, to correlate the evolutionary theory with any crying social need of DARWIN's day. Even in applied science the cause-and-effect relationship frequently fails to work. CHRISTOPHER COLUMBUS stood in desperate want of a steamship when undertaking his famous voyage; he would have been still better served by an airplane. The Indians could have saved their continent from the white man by developing the atomic bomb.

The point is not merely that some ages and peoples are more mechanically minded than others, but, more important, that most great inventions rest upon a cumulative series of subsidiary inventions. Technological discovery proceeds by progressive steps, and it seldom occurs that a single individual can skip several of these stages to achieve the final goal. Once the stationary steam engine was perfected, it was possible to plan to hitch it to a boat, but COLUMBUS's contemporaries could not reasonably have taken this giant leap into the future. Even with the prerequisite knowledge available, the urgency of a want may not produce the looked-for results. America's vast distances clearly called for the steam locomotive, but actually it was the "tight little island" of Britain that invented it. In this case, however, we know that JOHN STEVENS and other Americans were already struggling with the problem and were on the point of succeeding when GEORGE STEPHENSON anticipated them.

It is also true that motives not related to social need have sometimes sired inventions. There are cases on record of mechanisms born long before their time, which later had to be reinvented. Here individual genius operated according to its own internal logic. In other instances high-powered advertising stimulated a public demand which would not otherwise have existed. It might almost be said in such cases that invention is the mother of necessity. In recent years the profit motive has operated in this regard with increasing effect and has been a potent makeweight for technological progress.

Like other monistic interpretations in history, the concept of social compulsion as the wellspring of technological change must be used with caution, but, when so applied, it is a rewarding tool of analysis. It not only accounts satisfactorily for the significant goals of invention, but also goes far toward explaining the phenomenon, already referred to, of different minds working independently at the same time on the same invention. Though England lacked America's reasons for a steam railway, she had urgent reasons of her own, notably the need to transport cheaply and quickly the bulky commodities of her factories and mines. A later and more striking case of parallel effort in the two countries is the separate discovery of the so-called Bessemer process of making steel. Still others are the steamboat and the dual discovery of the principle of electromagnetic induction.

Within the United States itself the outstanding illustrations include the reaper and the telephone, possibly also the telegraph, the sewing machine, and the airplane. These occasions of multiple discovery would seem to constitute proof positive of the workings of social necessity; yet in some equally noteworthy instances the connection is far from clear. Why, for example, did a Boston dentist and a rural Georgia physician independently originate the use of ether as an anesthetic? What special conditions in the 1840's, common to New England and the Deep South, necessitated this conquest of pain? Such apparent exceptions the careful historian must take into account, but they should not shake his faith in the usual relationship of events.

In considering the impact of society on science, it should not be forgotten that science acts with reciprocal impact on society. A stock example of history textbooks is that of the cotton gin. This instrument, by rendering cotton culture profitable in the interior South, rejuvenated the dying institution of slavery, ensured its extension throughout the section, and thereby set the stage for the Civil War. Another favorite instance is the part played by the railroad in speeding Western settlement and, especially, its role in binding East and West by ties of mutual interest in time for them to present a united front to Southern secession. These two illustrations, when placed in juxtaposition, are instructive. In the first instance science functioned as a divisive factor politically; in the second, as a nationalizing factor. In neither case was the remoter outcome envisaged when the mechanism was devised. The inventor in accomplishing an immediate purpose loosed a stream of influences that ramified in unpredictable directions. An impersonal force, science operates on human affairs with unforeseen and often unforeseeable effect. For this if for no other reason it challenges the historian's most thoughtful attention.

The point may be further underscored by reference to some less obvious examples. CHARLES GOODYEAR's discovery of the secret of vulcanized rubber prepared the way for numerous unexpected applications. On the one hand, the new substance provided the necessary insulating material for the future age of electricity and also the tires which would hasten the coming of the bicycle and the automobile. On the other hand, vulcanized rubber had a profound effect on human health, ranging in use from the nipple on the infant's nursing bottle to the hot-water bag of the aged invalid. In the form of raincoats and overshoes it offered protection against common colds and rheumatism as well as against more fatal disorders such as pneumonia and influenza. Undoubtedly it was an important element in increasing the life span of the American people. Yet so unschooled are historians of science in reckoning with influences outside their narrow confines that no treatise on medicine has yet admitted CHARLES GOODYEAR to the roll of great healers of mankind.

A final example may well center on the lengthening span of human existence, to which allusion has just been made. Without discussing the many factors that were responsible, the promise of life at birth increased from around 35 years in 1789 to about 40 years in 1855 and to approximately 65 years at the present time. Nobody, so far as I know, has ever attempted to translate these juiceless abstractions into their social and intellectual significance. With an expected stay on earth of but 35 or 40 years, a man saw time in a very different perspective from today. By 18 or 20 he had lived half his life and should have attained the same relative position in the world as a person of 32 or 33 at present. This circumstance sheds light on the fabulous accomplishments of young people in those days: the youthful captains of merchant vessels and privateers; boys and girls marrying in their teens; GEORGE WASHINGTON holding public office as county surveyor at the age of 17; ALEXANDER HAMILTON addressing a patriot mass meeting at 17 and becoming General WASHINGTON's aide at 20; DAVID G. FARRAGUT starting as midshipman before the age of ten. One sees that the Framers of the Constitution acted with characteristic conservatism in restricting membership in the House and Senate to persons of 25 and 30 and reserving the presidency for patriarchs of 35.

But the whole time scale changed with the progressive prolongation of life after the Civil War. Youth needed no longer to shoulder the burdens of manhood. Boys now had ample leisure for play and slow maturing, for getting 8 or 12 years of education, even in many cases for going to college and spending additional time in studying for a profession. Child labor fell increasingly into disrepute, while legislative provision for child welfare became a major concern of society. It would take someone wiser than I to decide whether the modern accent on youth has not involved losses as well as gains, but there can be no doubt that there has resulted a significant reorientation of American life.

History, if written with a due appreciation of the role of science and technology, should have the useful effect of bridging the gap which in academic life too often separates those who study Nature from those who study human nature and society. If a scientific colleague of mine is to be believed, the college curriculum consists of the natural and unnatural sciences. If the person to whom he addressed this remark is to be believed, the distinction lies between the social and antisocial sciences. In support of this latter view a recent writer on higher education maintains that in the modern university the laboratory specialist alone lives in an ivory tower, indifferent to the tumult of events and a consistent foe of reform and change. To the extent that this may be true the layman is puzzled since the very essence of science and technology is experiment and growth. The answer, of course, lies in the fact that to the scientist's way of thinking social experimentation is not experiment at all but a misapplied figure of speech. A social experiment takes place under uncontrolled conditions, and because of the intrusion of numerous imponderables the results cannot be verified by repetition. Long ago ARISTOTLE realized that science in this exacting sense cannot be applied to human affairs, that on the contrary such studies as history, economics and sociology rest necessarily upon thoughtful observation.

Nevertheless the scientist, like his lay fellows, lives in a world where to stand still is to move backward, where intolerance of governmental and social change may invite disaster. And the scientist himself has been perhaps the chief contributor to this dizzy tempo of development. Can he then safely step aside and let the forces he has invoked work out their will blindly? Must he instinctively reject the fumbling, trial-and-error efforts of government to adjust society to the advances of science and technology? While not all scientists have been so minded, the examples of those who have are many and conspicuous. The historian by offering a truer picture of the past place of science may help to correct their vision. The invention of the atomic bomb seems also to be contributing to the same end. The international band of chemists, physicists, and engineers who created this terrible engine of destruction can hardly dodge the responsibility of trying to shape the kind of society that can be trusted with its use. This attitude may in turn foster a keener awareness of the implications of science for other affairs of life. In the end, we may hope, the natural and social sciences, each contributing to the wisdom of the other, will stand united in forwarding the common welfare of this nation and the world.

THE NEED FOR STUDIES IN THE HISTORY OF AMERICAN SCIENCE

By Richard H. Shryock

It is a truism that no factor has been more important, in the evolution of contemporary society, than the natural sciences and their technological applications. In many ways, science has been the most dynamic influence in making our present culture—in both its material and immaterial aspects—so different from that of preceding centuries. The history of the sciences, therefore, deserves emphasis in all accounts of the development of modern civilization.

Unfortunately, there is rarely any such emphasis in general historical narratives. Recent accounts of this nature have usually been written, in the United States, by professional historians. Prior to about 1900, these scholars stressed economic or political themes. Since then they have gradually given more attention to social and cultural developments; but science, in comparison with social movements, literature, and the fine arts, has fared badly even in this so-called "New History." Few professional historians have attempted studies of any given subject like physics or chemistry; and their discussions of science in general have usually been inadequate. Often these have been too brief, when present at all. A recent and otherwise superior text on "American History since 1914" gives as much as three pages to the work of a single literary figure like Sherwood Anderson, but less than a paragraph to the whole story of pure science in this period. In other cases, the comment fails to convey any real understanding of the problems or trends in the field noted.

The explanation of this is obvious enough. The training of historians continued to be largely political or literary in nature, long after they had transcended political themes and had ceased to be especially concerned with literary skills. Lacking training in science, even those who recognized its significance tended to avoid so unfamiliar a theme. Or, if they felt bound to include it, the discussion was likely to be of a superficial nature. Literature and the fine arts fared better, because here the scholar felt more at home.

The scientist would probably agree with these statements, and—if he were concerned at all—be inclined to blame historians for not having taken the training necessary for the interpretation of technical fields. Yet the historian could just as well blame the scientist. The former can hold that general history is not a subject in itself, but rather a synthesis of all phases of the past. And these specific themes should be handled by specialists. Why have the chemists not prepared more adequate histories of chemistry? If these were available, then professional historians could incorporate the findings in their own narratives. It may be added, incidentally, that the same reply could be made to social scientists, who sometimes bewail the lack of studies in social history by historians. It is the latter who may claim the grievance: why have sociologists not traced the story of the family that historians need for their syntheses?

Once again, the explanation is obvious. The scientist knows his technical field but is usually unfamiliar with historical sources, methods, background, points of view. His writings are apt to be of an antiquarian character and to lack unity, emphasis and coherence—simple as these literary qualities may seem at first glance. Thus, Garrison's *History of Medicine* is so disjointed that it can hardly claim to be a narrative, though it is valuable as a reference work. Even if an individual scientist happens to possess literary skill, his account of a particular field often is not related to the social background. And science can be no more understood out of relation to its total setting than the latter can be without reference to science.

In other words, if a chemist or biologist essays to write the history of his field, he needs historical training, just as a historian who wishes to deal with the same theme requires some degree of technical understanding. There is no need to decide which of these two types employs the best approach—the scientist or the historian. It may be that, in terms of their respective backgrounds, the scientist will be at his best in relatively technical accounts, the professional historian in the broader or more generalized interpretations. But there is work aplenty for both, and cooperation is clearly indicated.

∴

All that has been said applies with especial force to American history. Conversely, more exceptions could be claimed in the case of writings on European history. In the latter sphere, we do have in English such standard works on the sciences as those of SARTON and of THORNDIKE, and a number of excellent studies of specific fields or periods.

Far different is the situation in American history. Such studies as those mentioned rarely give much heed to the American fringe of European culture. Even "U-S-A-sian" scholars or scientists writing in this field have rarely dealt with the story of their own country. There is, of course, a cumulative literature on the history of American science, but most of this is in the nature of articles, biographies, and collections of short essays. With a very few exceptions, comprehensive studies—similar to the many works on other phases of national development—have never appeared. Where, for example, are treatments of American science as a whole comparable to KIRKLAND'S *History of American Economic Life,* or the *Cambridge History of American Literature?* Or the studies of special sciences, comparable to the innumerable works on special phases of our economic, political, religious, or literary evolution?

It may be that bibliographical checks would reveal a number of works on given sciences which are not well known, for these sometimes turn up unexpectedly. A history of American anthropology was published ten years ago, of all places in Calcutta. But a study that is little known hardly meets the needs here under consideration.

It is easy to see why histories of science have usually ignored national boundaries, whether of this or any other country. Each sovereign state necessarily has its own political, and to some extent its own economic story, but this is not so clear in the case of science. The latter has always maintained international ideals in large degree; and a nationalistic approach to its history might distort both the facts and the whole spirit of true science. It is quite plausible to conclude that American activities should be noted only when these emerged on the level of international significance.

But there is another side to this picture. If the history of scientific work done in this country is ignored, save at the points where it occasionally attracted international attention, much that was significant for the story of the American people is lost. And we certainly wish to trace the cultural development of our people as a whole. In a word, the value of studies in the history of American science is not to be found primarily in contributions to the history of science as such, but rather to the history of the United States.

There is a close parallel here to the current interest in other phases of our national culture; for example, to the history of American literature. Many scholars working in this field feel that their labors are directed toward enriching our knowledge of national life rather than toward a study of art for its own sake. Another parallel may be found in research on the history of American religion. Religious ideals are presumably international in nature and the great landmarks in religious history have not graced the American horizon; yet no one doubts the importance of integrating American religious history with other phases of national development.

. . .

Once it is agreed that the study of the history of science within any particular country is essential to the larger story of that country, further observations may be submitted which guard against the limitations of the nationalistic approach. First, writers on American—or British, or French—science should consciously guard against nationalistic prejudice. Second, they should "keep an eye out" for evidence in their respective territories of any phenomena of international significance that may hitherto have been overlooked. These may occasionally relate to a particular person, idea, or discovery. In the case of American science, to be specific, more adequate accounts might do something to achieve greater recognition abroad. To the extent that American work may have been discounted because of disdain for a colonial culture, such recognition may at times be really deserved. There have been individual Americans like BEAUMONT or WILLARD GIBBS, who were more appreciated in Europe than at home; but by and on the large, Europeans know as little of our scientific as of other phases of our national past. I once enjoyed a discussion with an able German medical historian who could not recall a single American physician prior to OSLER, although he was able to check in his library on BENJAMIN RUSH, when that gentleman was mentioned. It was presumably to remedy this situation that SIGERIST published at Leipzig, during the 1930's, his *Amerika und die Medizin;* but such interpretive efforts in Europe have been few and far between.

In other cases, a knowledge of American experience might throw new light on that of the "old countries"—by way of contrast. Thus an understanding of the handicaps suffered by scientists here because of a lack of certain attitudes or institutions may lead to new evaluations of the latter as these did actually develop overseas. In these ways, a properly conducted study of national history may, as a sort of by-product, improve our understanding of even the international record.

. . .

In reviewing the actual history of American science, one is often struck by the lag in research in this country as compared with the record of western Europe. This can hardly be explained by the lack of sufficient population and wealth to support pure science in a new land, since American resources in this respect were more than adequate a century ago. No doubt various factors explain the contrast, but one major influence seems to have been the prevailing attitude of the American people towards science and scholarship as such. Research lacked the prestige in this country that was accorded it in most European states, and consequently also lacked until recently the essential support of individuals, corporations, or governments. It is a question whether, even down to the present War, pure science—or pure scholarship for that matter—received the general respect in the U. S. A. that has long been obvious in such countries as Germany and Russia. Here the practical inventor like EDISON was widely respected, while WILLARD GIBBS was largely ignored.

This American attitude, whatever its historical explanations, is deeply imbedded in our national tradition and even in our contemporary culture. It was illustrated, until recently, by contrasting public attitudes toward the "professor" in Germany and in the United States. It is also reflected in the very lack of interest in the history of science in this country, which is the subject of the present statement.

Conversely, it follows that *one way to overcome American indifference to research is to give more attention to its history*. As special studies on the history of American science appear, something of their contents will percolate into college texts on our national story, and from these both facts and viewpoints may eventually reach the high school literature. In due time, a generation that knows something of GIBBS as well as of EDISON may be expected to support pure science in a manner hitherto lacking in our society. This educational process will afford an essential background to whatever particular measures are taken by either private or public institutions to further scientific advance. In other words, we must work out in a democratic society a public appreciation of pure science, in the place of the support long given it by a more or less aristocratic tradition in Europe.

It is not intended to imply, here, that historical studies should be primarily motivated by practical goals. On the other hand, such contemporary applications as are suggested above in the case of the encouragement of research, are certainly to be welcomed as by-products of historical scholarship.

In view of all these circumstances, it is seriously urged that a number of works be planned to trace the history of the major natural sciences and their technological applications in this country. Not every field would necessarily be undertaken at once. In a few cases, present treatments may be considered adequate; or, again, it may be impossible for some time to find a qualified author who can give the necessary time. It might even prove desirable, first, to experiment modestly with studies in only two or three subjects, especially if the right authors seemed available. But an over-all plan should be worked out looking toward an ultimate treatment of all the chief fields.

In the course of preparing such a plan, various questions concerning subjects, authorship, publication, etc., would inevitably arise. Just what is available on the American story? Should disciplines overlapping the natural and social sciences (geography, anthropology) be included? Would co-authorship be helpful? Should there be a final volume of synthesis? Ought there to be a formal series, brought out in uniform manner by a single publisher? It is unnecessary to answer such queries in advance of actual planning.

A word should be added, however, as to the general standards which should be approximated in the history of any given science. This should ordinarily cover a considerable part of the nation's past (let us say, at least a century), and of the nation's territory. Studies of a more limited scope might well have special values, but would partake of the nature of monographs and would not fully meet the purpose here in view. An adequate history would be interpretive as well as factual. It ought not to bog down in biographical details—a tendency still characteristic of much writing on the history of science—and would in general exhibit the forest as well as the trees. The history of thought should be treated, as well as that of institutions. The story should be presented against both the social and the intellectual background of a given time and place. And, as far as possible within limits imposed by the nature of the materials, the treatment should be intelligible to the serious lay reader.

In conclusion, one very natural objection to these suggestions may be anticipated. It is sometimes held that books should not be planned in series; should not be made to order. One should rather wait for inspiration to strike—laissez faire. No doubt the product of an individual's spontaneous enthusiasm is apt—other things being equal—to be superior to one solicited by a planning board. But there may be times when individuals possessing some potential enthusiasm require encouragement, or need to have their interests correlated with those of similar colleagues, and then a planning group may be of real service. Presumably the latter would never urge an author to undertake a work if he had not already displayed some inclination in that direction. At any rate, in the present instance spontaneous individual interest has failed to produce—decade after decade—the needed works. One may conclude, therefore, that a laissez faire procedure has proved inadequate and that a planned historical program is in order.

FRANKLIN'S EXPERIMENTS ON HEAT ABSORPTION AS A FUNCTION OF COLOR

By I. Bernard Cohen

I

Of all Franklin's non-electrical experiments, those which he conducted with a view to determining the relation of color of material to heat absorption and thermal conductivity are the best known. And, incidentally, these have been most subject to criticism.

The usual source of information concerning these experiments is a letter which Franklin wrote to Miss Mary Stevenson[1] on September 20, 1761. This letter was introduced by Franklin into the fourth edition of his *Experiments and Observations on Electricity*[2] and was reprinted in the fifth edition. The portion relating to these experiments reads as follows:

As to our other Subject, the different Degrees of Heat imbibed from the Sun's Rays by Cloths of different Colours, since I cannot find the Notes of my Experiment to send you, I must give it as well as I can from Memory.

But first let me mention an Experiment you may easily make yourself. Walk but a quarter of an Hour in your Garden when the Sun shines, with a part of your Dress white, and a Part black; then apply your Hand to them alternately, and you will find a very great Difference in their Warmth. The Black will be quite hot to the Touch, the White still cool.

Another. Try to fire Paper with a burning Glass. If it is White, you will not easily burn it; but if you bring the Focus to a black Spot, or upon Letters, written or printed, the Paper will be immediately on Fire under the Letters.

Thus Fullers and Dyers find black Cloths, of equal Thickness with white ones, and hung out equally wet, dry in the Sun much sooner than the white, being more readily heated by the Sun's Rays. It is the same before a Fire; the Heat of which sooner penetrates black Stockings than white ones, and so is apt sooner to burn a Man's Shins. Also Beer much sooner warms in a black Mug set before the Fire, than in a white one, or in a bright Silver Tankard.

My Experiment was this. I took a number of little square Pieces of Broad Cloth from a Taylor's Pattern-Card, of various Colours. There were Black, deep Blue, lighter Blue, Green, Purple, Red, Yellow, White, and other Colours, or Shades of Colours. I laid them all out upon the Snow in a bright Sunshiny Morning. In a few Hours (I cannot now be exact as to the Time), the Black, being warm'd most by the Sun, was sunk so low as to be below the Stroke of the Sun's Rays; the dark Blue almost as low, the lighter blue not quite so much as the dark, the other colours less as they were lighter; and the quite White remain'd on the Surface of the Snow, not having entred [*sic*] it at all.

What signifies Philosophy that does not apply to some Use? May we not learn from hence, that black Clothes are not so fit to wear in a hot Sunny Climate or Season, as white ones; because in such Cloaths the Body is more heated by the Sun when we walk abroad, and are at the same time heated by the Exercise, which double Heat is apt to bring on putrid dangerous Fevers? That Soldiers and Seamen, who must march and labour in the Sun, should in the East or West Indies have an Uniform of white? That Summer Hats, for Men or Women, should be white, as repelling that Heat which gives Headachs to many, and to some the fatal Stroke that the French call the *Coup de Soleil?* That the Ladies' Summer Hats, however, should be lined with Black, as not reverberating on their Faces those Rays which are reflected upwards from the Earth or Water? That the putting a white Cap of Paper or Linnen *within* the Crown of a black Hat, as some do, will not keep out the Heat, tho' it would if placed *without?* That Fruit-Walls being black'd may receive so much Heat from the Sun in the Daytime, as to continue warm in some degree thro' the Night, and thereby preserve the Fruit from Frosts, or forward its Growth? — with sundry other particulars of less or greater Importance, that will occur from time to time to attentive Minds? I am

Yours affectionately,

B. Franklin.[3]

Until now, this has been the sole source of information concerning Franklin's experiments. No one has discovered the original "Notes" of the "Experiment," to which Franklin refers in the first paragraph quoted above. Nor do his other letters or his "Autobiography" give any clue as to when the experiment was made, whether in the 1750's, the 1740's, or even in the 1730's.

Mr. Carl Van Doren called my attention to an item in the "Miscellaneous Papers" in the Franklin Collection of The American Philosophical Society Library, described in the Calendar[4] as follows:

By ——— ———. 1736–7. January 25.
Experiments with various colors to show which imbibe and which repel the sun's rays. Diss. 1 p.

From the description, it seemed at once that this was the page of "Notes" referred to by Frank-

[1] Mary Stevenson later became Mrs. Hewson. She is also known as "Polly" Stevenson [or Hewson]. An account of Franklin's relationship with her may be found in Carl Van Doren: *Benjamin Franklin* (New York, the Viking Press, 1938).

[2] The fourth edition was printed in London in 1769, the fifth in London in 1774. A new edition of Franklin's book on electricity has just been printed, edited with an introduction by the present writer: *Benjamin Franklin's Experiments, a new edition of Franklin's "Experiments and Observations on Electricity"* (Cambridge, Mass., Harvard University Press, 1941). Chapter IV of the Introduction contains an account of the various eighteenth-century editions of Franklin's book.

[3] The portion of the letter which I have printed above follows the version printed by A. H. Smyth [editor]: *The Writings of Benjamin Franklin* (10 vols.; New York, The Macmillan Co., 1905–07), Vol. IV, pp. 111 ff., since Smyth's version is a transcript of the original letter, then in the possession of T. Hewson Bradford, M.D., a descendant of Mary Stevenson's husband, Dr. William Hewson.

[4] I. Minis Hays [editor]: *Calendar of the Papers of Benjamin Franklin in the Library of the American Philosophical Society* (5 vols.; Philadelphia, The American Philosophical Society, 1908), Vol. IV, p. 173.

LIN. And, although not written in FRANKLIN's own hand, it might very well have been a copy that he had made in order to send it on to someone else, perhaps even to Miss STEVENSON. This was a practice common enough with FRANKLIN.[5]

This document reads as follows:

Jan[ry] 25th 1736/7.

EXPERIMENTS with Colours of various Sorts, to shew which imbibe, and which repel, or do not readily admit, the Sun's Rays; made with placing on the level Snow, Bitts of Linnen, Silk, Leather, Paper, Woollen Cloth, Feathers, & other Materials; exemplify'd by six Degrees of melting down or sinking in the Snow. By which it was observable that the different Weights or kinds of the Subjects made no Alteration; all the Variety of Effects being owing only to the Difference of Colours, except the piece of Glass.

1st Degree — Shallowest. White.
2nd ——— Less shallow. Light Red. Red & White striped. Light Yellow. Light Azure.
3rd ——— Least shallow, meaning next above deep. Lively Blood Red. Reddish Brown, or Bright Cinnamon.
4th ——— Deep. Deep Grass Green. Yellow Brown, or dirty Yellow.
5th ——— Deeper. Deep Blue. Gloomy Red. Dark Olive, or dark Brown.
6th ——— Deepest. Black. Also a piece of Window Glass.

Note. Some of the above Colours may be a little misnamed; or if another Man should make the Use of such Colours as he would take for some of the above mentioned, his Experiments might not exactly agree with these, especially in those that are nearest the Middle between the two Extreams of Black and White; therefore it may be proper to note for a Certainty that such Things as are perfectly Black, sink deepest; such as are dark Brown, dark Blue, or otherwise nearest akin to Black, sink the next deepest. And such Things as are most perfectly White sink the least, and such as are of a pale Flame Colour, or that vary the least from Witeness and Light, sink the least next to White.

Mem°. An easy Way to make a Couple of Experiments that will be very evident and give Satisfaction, is to take an Inch or two of the Feather end of a black Quill and the like Part of a white Quill; also to black with Ink one End of a Slip of white Paper, and the other End remain white; put these on the Snow where the Sun comes, and in an Hour's Time the Black of each will be sunk below the surface (perhaps ½ an Inch, as the Heat may be) and the White scarce visibly settled nor will the White settle so much in a whole Day.

An Observation may be added That those Things which are most closely woven, or are least porous, do most readily shew the Sun's Effect upon them. Also such as have one Side black and the other Side of a different Colour, will imbibe more Heat with the black Side next to the Sun.[6]

This record of experimentation is not, however, a note of FRANKLIN's experiment at all. It is a transcript of an experiment made by FRANKLIN's friend, JOSEPH BREINTNAL. But, since this transcript appears to have been in FRANKLIN's possession, one may assume that early in 1737 FRANKLIN knew of BREINTNAL's experimental proof of the effect of color on thermal absorption and conduction. If FRANKLIN actually made this experiment himself, and if he did so independently of BREINTNAL, then clearly his own experiment must have been made prior to 1737. That this is the case will be seen from an examination of BREINTNAL's papers, in which we find the original account, from which the above transcription was made.

II

JOSEPH BREINTNAL[7] (or BREINTNALL) [*died* 1746] was a friend of FRANKLIN's, an original member of the Junto (the precursor of the American Philosophical Society). BREINTNAL was associated with FRANKLIN in publishing "The Busy-Body" in ANDREW BRADFORD's newspaper, the *American Weekly Mercury,* and assisted FRANKLIN in various ways when he entered business.[8] For a while BREINTNAL was Sheriff of Philadelphia county and was Secretary of the Library Company of Philadelphia from the date of its organization in 1731 until BREINTNAL's death in 1746, when FRANKLIN himself took over the office. As Secretary of the Library Company, BREINTNAL became a correspondent of the library's London agent, PETER COLLINSON, who was chiefly responsible for FRANKLIN's initial interest in electrical research. BREINTNAL arranged for JOHN BARTRAM to gather plants for COLLINSON, and thus started a famous correspondence between the patron of American science and the Pennsylvania farmer who came to be, as LINNAEUS supposedly remarked, "the greatest living natural botanist."[9]

BREINTNAL's correspondence with COLLINSON reveals a keen interest in natural history and one letter, in which BREINTNAL describes an Aurora Borealis, was introduced by COLLINSON into one of the meetings of the Royal Society of London and then printed in the journal published by the Society.[10] In addition, he conducted a remarkable series of experiments in order to determine some of the properties of heat. Not that these experiments have any remarkable qualities when viewed today; for such is not the case. They might today be performed by any schoolboy. But, at the time

[5] An outstanding example of this practice is the copy of his letters on electricity which FRANKLIN had copied in 1750 for JAMES BOWDOIN. These are described in detail in Chapter IV, pt. 2, of the Introduction to *Benjamin Franklin's Experiments* (cited in footnote 2 *supra*).

[6] Published from the original in the American Philosophical Society Library: *Franklin Papers,* XLIX, 37.

[7] Cf. STANLEY BLOORE: "Joseph Breintnall, First Secretary of the Library Company," *Pennsylvania Magazine of History and Biography* LIX (1935), pp. 42–56.

[8] The various aspects of FRANKLIN's relations with BREINTNAL are recorded in FRANKLIN's *"Autobiography."* Cf. VAN DOREN: *Benjamin Franklin, (op. cit).*

[9] Cf. WILLIAM DARLINGTON: *Memorials of John Bartram and Humphry Marshall* (Philadelphia, Lindsay and Blakiston, 1849), esp. p. 112.

[10] *Phil. Trans.* XLI (for 1739 & 1740), p. 359. A second letter concerning a rattlesnake bite was read at a meeting of the Royal Society after BREINTNAL's death and is published in *Phil. Trans.* XLIV (for the year 1746), pp. 147–50.

when they were made, there were not many people in the American Colonies who were carrying out original experiments in physics.

The most important of BREINTNAL's experiments concerns the penetration of the calorific effects of the sun's rays upon pieces of differently colored cloth placed on the snow. His memorandum concerning the experiment reads *in extenso* as follows:

Augst 3rd 1737.

That the Heat of the Sun penetrates such Things as are colour'd more than such as are White, will appear by the following Experiments, which I was induced to make about seven Years ago, and lately to repeat, from some Hints given me by Benjn Franklin, and from observing that People who come among us from the warm Islands do most of them wear whitish Cloths, in which I suppose they find themselves cooler than in Others tho' few of them may know the Reason of it; from observing also that a small Glass will not burn white Paper, tho' it easily does it if the Paper be stain'd, and from taking Notice of a young Woman's complaining that her black Gloves had burnt her Hands, &c.

January 25. 1736.

Experiments with Colours of various Sorts, to shew which imbibe, and which repel (or do not readily admit) the Sun's Rays; made with placing on the level Snow, Bitts of Linnen, Silks, Leather, Paper, Woollen Cloth, Feathers, and other Materials — exemplify'd by six Degrees of melting down or sinking in the Snow. By which it was observable that the different Weights or Kinds of the Subjects made no Alteration; All the Variety of Effects being owing only to Difference of Colours, except the Piece of Glass, and except the Circumstances of Cloth being closely or loosely wove —

1st Degree.	Shallowest	White.
2.	Less shallow	Light Red. Red & White striped. Light Yellow. Light Azure.
3.	Least shallow, meaning next above deep	Lively Blood Red. Reddish Brown, or bright Cinnamon.
4.	Deep	Deep Grass Green. Yellow Brown, or dirty Yellow.
5.	Deeper	Deep Blue. Gloomy Red. Dark Olive, or Dark Brown.
6.	Deepest	Black. Also a piece of Window Glass.

Note. Some of the above Colours may be a little misnamed; or if another Man should make Use of such Colours as he would take for some of the above mentioned, his Experiments might not exactly agree with these, especially in those that are nearest the Middle between the two Extreams of Black & White; Therefore it may be proper to note for a Certainty, that such Things as are perfectly Black sink deepest; such as are dark Brown, dark Blue, or otherwise nearest akin to Black, sink the next deepest. And such Things as are most perfectly White sink the least, and such as are of a pale Flame Colour, or that vary the least from Whiteness and Light, sink the next least to White.

And those Things which are the most closely woven, or are least porous, do most readily shew the Sun's Effect upon them. Also such as have one Side black and the other Side of a different Colour, will imbibe more Heat with the black Side next to the Sun.

An easy Way to make a Couple of Experiments that will be very evident & give Satisfaction, is to take an Inch or two of the Feather End of a black Quill; and the like Part of a white Quill, also to black with Ink one End of a Slip of white Paper, and let the other End remain white; put these on the Snow where the Sun comes, and in an Hour's Time the Black of each will be sunk below the Surface (perhaps ½ an Inch, as the Heat may be) and the White scarce visibly settled; nor will the White settle so much in a whole Day.[11]

The similarity of this document and the one in FRANKLIN's possession (published in the preceding section) leaves no possible doubt as to their identity of source. The experiment is dated January 25, 1736/7 in the first and simply January 25, 1736 in the second. This means that the date is actually January 25, 1737 (Julian) in the "new style" of reckoning years from January 1 rather than the older English style of reckoning years from March 25.[12] The document last quoted is in BREINTNAL's hand, but the other seems in a different hand. Most likely BREINTNAL had a copy made for FRANKLIN and this copy does not contain the preamble which BREINTNAL kept for the sake of the record, but merely contains an account of the experiment proper. Although the material contained in both is identical, item for item, the style varies in several details as may be seen by comparing them.

Mr. BLOORE, in an article on BREINTNAL, would have us believe: "The memorandum recording the information is dated August 3, 1737, but the experiment had been originally performed seven years previously. It was repeated at the later date as the result of hints given by FRANKLIN."[13] But the text of the memorandum itself is hardly a sufficient warrant for this inference, which would make BREINTNAL the originator of the idea. The memorandum states simply: "the following Experiments, which I was induced to make about seven Years ago, and lately to repeat, from some Hints given me by Benjn Franklin . . . ." Since the words "and lately to repeat" form a parenthetical expression, it is equally likely that the memorandum implies that BREINTNAL was induced to make the experiments "about seven Years ago . . . from some Hints given me by Benjn Franklin." He might, or equally well might not, have been induced to repeat the experiments from hints of FRANKLIN or from the other observations listed in the preamble.

The date when BREINTNAL made the original

[11] Published from the original in the Historical Society of Pennsylvania: *Logan Papers,* X, 100.

[12] The English calendar was reformed in 1752, at which time the older Julian calendar was abandoned and the Gregorian calendar accepted. At this time, too, the beginning of the year was set at January 1st, rather than at March 25th which had previously marked the beginning of the year. During the century or so prior to this reform, most writers used the form "1751/2" or "1751,2" for dates between January 1 and March 25, in order to avoid confusion.

[13] *Loc. cit.*

experiments was "about seven Years ago," or some time around 1729. This dates the experiment soon after the organization of the Junto. BREINTNAL was a charter member of the Junto and FRANKLIN recommended that at the meetings,

> . . . Mr. Breintnal's poem on the Junto be read once a month, and hummed in concert by as many as can hum it.[14]

FRANKLIN, who dominated the club, wanted its members to discuss literary, philanthropic, philosophical, and scientific problems. Some of the more scientific questions raised for discussion were:

> Is *sound* an entity or body?
> How may the phenomena of vapours be explained?
> How may smoky chimneys be best cured?
> Why does the flame of a candle tend upwards in a spire?
> Whence comes the dew, that stands on the outside of a tankard that has cold water in it in the summer time?[15]

If BREINTNAL made his first experiment in the early years of the Junto as he himself declares, one may be as certain as can be that FRANKLIN would have known about it. This would have been the very sort of thing to be discussed at Junto meetings and BREINTNAL, as a good Junto member, would have been more than pleased to offer the results of his own experiments as a topic of "philosophical" discussion.

FRANKLIN's experiments and BREINTNAL's are not identical. According to FRANKLIN's letter to Miss STEVENSON, he used sample pieces of cloth from a tailor's card. BREINTNAL used pieces of cloth and feathers and pieces of cardboard. FRANKLIN wrote to Miss STEVENSON that the experiment in question was one which he had *himself* performed. He did not write that it was a friend's experiment, but rather that he had performed the experiment himself. FRANKLIN was usually more than generous in giving credit to his friends and co-workers for their discoveries, especially in the fourth edition of his book on electricity, in which the letter to Miss STEVENSON was published. Thus he introduced footnotes into that edition (the first edition that he personally supervised) explaining that such and such a discovery had been made by SYNG, another by HOPKINSON, and others by KINNERSLEY.[16] Since he did not follow this procedure in the case of BREINTNAL, one must assume that the reason lies in the fact that the original idea was FRANKLIN's. This would support the reading of BREINTNAL's statement which I suggest above, namely, that BREINTNAL had been induced to make the experiments "about seven Years ago . . . from some Hints given me by Benjamin Franklin."

A logical picture emerges if we admit the following reconstruction.

1. Some time about 1729 or earlier, FRANKLIN was led to make the experiment with sample squares of cloth and discovered that the amount of heat absorbed from the sun's rays and conducted by the cloth was a function of the color.
2. Since FRANKLIN was interested in encouraging the pursuit of science, that is "experimental philosophy," among the members of the newly-founded Junto, he suggested to BREINTNAL (and possibly to other members) by means of "hints" that experiments on absorption of solar heat by variously colored substances might reveal some interesting information.
3. Acting on FRANKLIN's suggestion, BREINTNAL made some experiments and noted, as FRANKLIN had already discovered, the effect of color.
4. Seven years later, on January 25, 1736/7, either as a result of an additional prodding from FRANKLIN or simply on his own initiative, BREINTNAL repeated the experiments with greater care and wrote up an account of his findings. This account was rewritten by a copyist and given to FRANKLIN.

Following this reconstruction, the date of FRANKLIN's original experiment can be set around 1729 at the very latest.

Harvard University

[14] JARED SPARKS [editor]: *The Works of Benjamin Franklin* (revised edition, 10 vols.; Philadelphia, Childs & Peterson, no date), vol. II, p. 552.

[15] *Ibid.*, footnote on pp. 9–10.

[16] *Benjamin Franklin's Experiments* (*op. cit.*), p. 149.

Franklin, Boerhaave, Newton, Boyle & the Absorption of Heat in Relation to Color

By I. Bernard Cohen *

BENJAMIN FRANKLIN's experiments on heat absorption were performed by placing on snow some black and white objects and pieces of cloth dyed in several colors, and then observing the degree to which the snow was melted under each specimen. In an earlier article in *Isis*,[1] the dates of these experiments were discussed in relation to the published and manuscript versions. The printed account appeared for the first time in the fourth edition of Franklin's book on electricity (London, 1769)[2] as part of a letter to Miss Mary Stevenson dated 20 September 1761. In this letter Franklin discussed "the different degrees of heat imbibed from the sun's rays by cloths of different colours," and referred to "my experiments," but he did not claim that the discovery of this effect had been made by him: he said only that he had made an experiment on this subject. Two memoranda of the experiments exist, one by Franklin's fellow Junto member, Joseph Breintnal, the other written in an unidentified hand. The latter is dated 25 January 1736/7; the former 3 August 1737, containing an account of experiments performed on 26 January 1736 (1737 N.S.).[3] In the introductory remarks to Breintnal's memorandum of 3 August 1737, Breintnal said that these experiments he had been "induced to make" and "lately to repeat" from "some hints given me by Benjamin Franklin" and from observing that a woman's hands had been "burnt" by wearing black gloves, that in "the warm islands" people wear "whitish cloths," and that a small magnifying glass will burn stained paper but not white paper. Breintnal said that the original experiments had been made "about seven years ago."

Since Franklin described the experiment to Miss Stevenson as "my experiment," and since Breintnal said that Franklin had suggested that he do the experiment, it can be assumed that either Franklin performed the experiment and suggested that Breintnal repeat it, or that Franklin knew the result in advance of experimentation and devised the form of the experiment for Breintnal to perform. The actual experiment, in the two descriptions of Breintnal, produced "six degrees of melting down or sinking in the snow." He used "bits of linnen, silk, leather, paper, woollen cloth, feathers, & other materials." In

* Harvard University.

[1] Franklin's experiments on heat absorption as a function of color, *Isis*, 1943, *34*: 404–407.

[2] These editions are described in *Benjamin Franklin's experiments: a new edition of Franklin's "Experiments and observations on electricity"* (Cambridge: Harvard University Press, 1941), intro., ch. iv.

[3] Both documents are printed in extenso in ref. 1.

Franklin's description of "my experiment" it is said that he "took a number of little square pieces of broad cloth from a taylor's pattern-card, of various colours . . . ." Since the materials used by the two experimenters were not identical, it is reasonable to conclude that both Franklin and Breintnal performed an experiment. Before describing "my experiment," Franklin told Miss Stevenson that if she walked in the garden on a hot sunny day wearing a dress that was partly white and partly black, she would find "a great difference in their warmth," a statement similar to Breintnal's remarks about the black gloves. Like Breintnal, Franklin also referred to the difficulty of igniting white paper with a burning glass and the ease of doing so if "you bring the focus to a black spot or upon letters, written or printed."

Before describing the experiment to Miss Stevenson, Franklin said:

> Thus fullers and dyers find black cloths, of equal thickness with white ones, and hung out equally wet, dry in the sun much sooner than the white, being more readily heated by the sun's rays. It is the same before a fire; the heat of which sooner penetrates black stockings than white ones, and so is apt sooner to burn a man's shins. Also beer much sooner warms in a black mug set before the fire, than in a white one, or in a bright silver tankard.

This statement so closely resembles some remarks of Boerhaave that there can be no question of the fact that Franklin's reading of the great Dutch master had influenced his thought. Boerhaave wrote:

> EXPERIMENT XIII
>
> 152. If this fire determined by the sun, be received on the blackest known bodies, its heat will be long retain'd therein; and hence such bodies are the soonest and the strongest heated by the same fire, as also the quickest dried, after having been moisten'd with water; and it may be added, that they also burn by much the readiest; all of which points are confirm'd by daily observations. Let a piece of cloth be hung in the air, open to the sun, one part of it dyed black, another part of a white colour, others of scarlet and diverse other colours; the black part will always be found to heat the most, and the quickest of all; and the others will each heat the more slowly, by how much they reflect the rays more strongly to the eye; thus the white will warm the slowest of them all, and next to that the red, and so of the rest in proportion, as their colour is brighter or weaker. This is well known to the nations who inhabit the hotter climates, where the outer garments, if of a white colour, are found best to preserve the body from the scorching sun; and black ones, on the contrary, to increase the heat. And it has often been observed by makers of woollen cloth, that if at the same time and place they hang out two wet pieces, the one black, the other white, the former will smoak and dry quickly, but the latter retain its water long; and cloths of other colours will dry so much the slower, by how much their colours are the brighter.
>
> 153. It has also been long observed, that all black bodies are sooner kindled and set on flame by the same fire, than those of any other colour. The dust of white touch-wood will hardly catch, and sustain a spark of fire struck upon it; whereas if the same be struck on a black coal, the dust hereof will readily receive, and keep it up, so that in a short time the whole dust will be on fire. The purest and whitest linen will hardly maintain a spark thrown upon it; but if the like spark be cast on tinder, which is only the coal of linen kindled, and again extinguished, it will immediately catch through the whole body of it. Nor would gun-powder, were it not for its black colour, be so easy to kindle; as appears by the powder made of white nitre ground with sulphur. The gardeners have long complain'd,

> that their white soils would not warm with the sun, except in the very outmost surface; whereas the black grows so hot, as to burn the roots of plants. The chemists have long ago observed, that black bodies, when committed to digestion, or reduced to blackness by art, easier grow hot by the same fire, in the *caput corvi, collum cygni,* or *cauda pavonis,* which require different degrees of fire. Lastly, the philosophers have confirmed the matter by experiments. If a piece of white paper be laid on the focus of a burning-glass, it will be long before it heat, and very long before it take fire; and as soon as kindled, quits its whiteness, turns brown, and then black; immediately after which it catches flame: whereas, if a black paper be laid on the same focus it immediately takes fire. We have some extraordinary things on this head in the experiments of the academy *del Cimento.* Hence we see the reason of many phaenomena in meteors; it being a known point, that thunder and lightning are never found more furious than in dark weather, when the heavens are cover'd with black clouds; from whence usually arise terrible whirlwinds, by the rarifaction of the air, occasion'd by the sudden immense production of heat.

This extract is taken from Peter Shaw's edition of *A new method of chemistry; including the history, theory, and practice of the art: translated from the original Latin of Dr. Boerhaave's Elementa chemiæ, as published by himself* (London: printed for T. Longman, 1741).[4] This work was called "the second edition," but the first edition, made by Peter Shaw and Ephraim Chambers (London 1727) had been based on a "spurious edition" originally printed in 1724 and prepared by Boerhaave's students, who had evidently collated their lecture notes and issued the book under their teacher's name. Boerhaave brought out a "genuine" version of his own book in a two-volume edition published in Leyden in 1732, which served as the basis for the Shaw "second edition" of 1741, and which was also translated into English by Timothy Dallowe in 1735.[5]

Franklin's letter to Miss Stevenson in 1761 shows great similarity to the text of the "genuine edition" of Boerhaave, which we know he had seen in Shaw's English translation. But the latter was published in 1741 and was based on the Leyden edition in Latin of 1732, which was reprinted in London in the same year. Hence the suggestion for the Franklin and Breintnal experiments must have come from the Latin version, since the experiments of Breintnal were dated 1737. Breintnal's phrase, "about seven years ago," indicates that he had no record of the exact date of the first experiments, nor did he recall just when they had been performed. Perhaps they had been done first "about five years ago" and not "seven."

Of course, Franklin may have read the text of the "spurious edition" in Latin of 1724 or the English version of it by Shaw and Chambers of 1727. Here Franklin would have found a briefer account of this subject, including only

[4] Vol. 1, pp. 262–263.

[5] See Tenney L. Davis, "The vicissitudes of Boerhaave's textbook of chemistry," *Isis*, 1928, *10*: 33–46; Menno Hertzberger, *Short-title catalogue of books written and edited by Hermann Boerhaave*, compiled with the assistance of E. J. van der Linden. Preface by J. G. de Lint (Amsterdam: Hertzberger, 1927).

The "first edition" of the Shaw and Chambers English version of Boerhaave's *Chemistry* was not based exclusively on the "spurious edition" printed in 1724 in "Paris" (actually Leiden), but also on various manuscript versions. Thus William Burton, writing about Boerhaave, thought this English version to be more reliable for Boerhaave's early views than the Latin one. Professor Henry Guerlac, who has kindly provided this information, discusses this point in his article on Lavoisier, soon to be published in *Isis*.

three colors: white, red, black. In this version, there was no mention of the preference for white clothes in hot weather, nor of the ignition of paper by focussing the sun's rays.[6] Had Franklin read this version of Boerhaave's *Chemistry*, and had he performed his experiments before encountering the "genuine version," he would be entitled to a claim of some originality in this matter — but only to the extent of enlarging the number of colors and the use of samples of tailor's cloths. In this case, the repetition of the experiment by Breintnal in 1737 would undoubtedly have been occasioned by Franklin's coming upon the "genuine edition" in Latin with its fuller discussion of this subject. In any event, long before Franklin published the account of his experiments, in the 1769 edition of his book on electricity, the phenomena of heat absorption in relation to color had been widely printed in the many editions and translations of Boerhaave's *Chemistry*.

Boerhaave's *Chemistry* studied by Franklin and so many other eighteenth-century physical scientists reveals an admiration for the British school of natural philosophy; there are references to the work of Hooke and Halley; Bacon is cited with veneration; Boyle is quoted again and again, always in terms of the highest admiration; respect is paid to Newton's *Opticks*.[7] Proposition VII of Book Two, Part III, of the *Opticks* was intended to show that "the bigness of the component parts of natural bodies may be conjectured by their colours." [8] This proposition is a mixture of observation and speculative hypothesis, in which Newton discussed the size of particles and their "order" in terms of their chemical action, and such physical characteristics as density. Thus Newton essayed a discussion of the "bigness of metallick particles," and in general the "thickness" of the particles of matter. Toward the end of this proposition, Newton indicated:

> . . . for the production of black, the corpuscles must be less than any of those which exhibit colours. For in all greater sizes there is too much light reflected to constitute this colour. But if they be supposed a little less than is requisite to reflect the white and very faint blue of the first order, they will . . . reflect so very little light as to appear intensely black . . . .

In this way we may understand why "fire, and the more subtile dissolver putre-

[6] The account of this subject in the "spurious edition" reads in extenso as follows: "Thus if you hang up several different coloured clothes, in a dark place; some white, others red, and others black; and place your self at a little distance; you will perceive nothing at all where the black are; where the white are, there will something appear; so altho where the red are, tho' less than where the white. Not that there is more fire in one than another; but one reflects more or fewer rays than another; so that the circumstances of light do not depend on fire, so much as on the surface of the body that reflects it." [42] H. Boerhaave, *A new method of chemistry; including the theory and practice of that art.* Translated by P. Shaw and E. Chambers. 2 vols. (London: J. Osborn & T. Longman, 1727), pp. 229–230.

[7] See Hélène Metzger, *Newton, Stahl, Boerhaave et la doctrine chimique* (Paris: Felix Alcan, 1930).

Although the text of Boerhaave's book abounds in expressions of admiration for the British school, citing Hooke, Halley, Bacon and Boyle, the most flattering references to these authors, including long extracts, are to be found in the footnotes written by Shaw. Despite Boerhaave's admiration for the British school, he did not follow them in their "kinetic" concept of heat, but advocated a "fluid" concept. This point is explored in greater detail in the writer's forthcoming monograph, Franklin & Newton, an inquiry into speculative Newtonian experimental science and Franklin's work in electricity as an illustration thereof — shortly to be published as a memoir of the American Philosophical Society.

[8] The quotations from Newton's *Opticks* are taken from the fourth edition, reprinted by Dover Publications, 1952.

faction, by dividing the particles of substance, turn them to black, why small quantities of black substances impart their colour very freely and intensely to other substances to which they are applied . . . ." Most important of all, Newton suggested that consideration of particle size affords an explanation as to "why black substances do soonest of all others become hot in the sun's light and burn (which effect may proceed partly from the multitude of refractions in a little room, and partly from the easy commotion of so very small corpuscles;) . . . ."

It should, therefore, be noted that the part of the discussion of the problem relating to the difference in absorption of solar heat between black and white substances, and the example of the ease with which the sun's light may ignite black material as compared to white, was posed by Newton. These questions entered Newton's discussion of the fundamental nature of matter, as related to the size of the ultimate particles or corpuscles of which bodies are composed. Since black and white were the extremes, and since the other colors ranged in order between these two extremes, the absorption of solar heat might well be expected to vary more or less continuously through the visible spectrum, but Newton did not say so. Boerhaave's extension of Newton's observations was ingenious, if not particularly astonishing; although it must be noted that Newton did not take up the question of the heating of bodies, but only the upper limit of heating — that is, actual ignition and burning.

Newton returned to this question in the "queries" at the end of Book III of the *Opticks*. In Query 5, he discussed bodies and light acting mutually upon one another in the sense that bodies emit, reflect, refract, and "inflect" (diffract) light and that light heats bodies and puts "their parts into a vibrating motion wherein heat consists." Newton asked, "Do not black bodies conceive heat more easily from light than those of other colours do, for the reason that the light falling on them is not reflected outwards, but enters the bodies, and is often reflected and refracted within them, until it be stifled and lost?" The "strength and vigor of the action between light and sulphureous bodies observed above" is probably, according to Newton's next query, one reason why "sulphureous bodies take fire more readily, and burn more vehemently than other bodies do."

Franklin and Boerhaave were both admirers of Robert Boyle and thus it is of interest to see that Boyle had also explored the problem of the heating of black and white materials by the sun's rays. In his *Experimental History of Colours*, Part II — "Of the Nature of Whiteness and Blackness," Boyle described how black and white tiles exposed to sunlight became heated at different rates. He also observed that an egg painted black and exposed to the sun cooked more quickly than a white egg, and that his hand would be warmed more rapidly by sunlight if encased in a thin black glove than in one of "thin but white leather." [9] Undoubtedly, similar observations had been made by others before Boyle.

Hence it may be seen that the experiments described by Franklin and by

[9] See J. F. Fulton, *A bibliography of the Honourable Robert Boyle, Fellow of the Royal Society*, Part VII, pp. 47–50. *The experimental history of colours* (1664), pp. 104, 126–127. The first edition was published in 1663. I am grateful to Miss Marie Boas for this reference.

Boerhaave belong to a tradition of explorations relating to heat and light that goes back well into the seventeenth century. Bacon had discussed the problems of heat and light in relation to solar radiation, and others had examined the possibility of light without heat, as in the case of moonlight. Later on, at the end of the eighteenth century, it was found that radiant heat might be transmitted without being accompanied by visible light.[10] Thus we can understand why, in his letter to Miss Stevenson, Franklin did not say, much less imply, that the effects of color on the absorption of solar heat had been a discovery of his; although some later writers have claimed that this should be included amongst his contributions to physics.[11] Always careful in scientific matters, Franklin merely said that he had performed an experiment, one that we see did differ from what had been reported by Boerhaave — if only in minor details. Perhaps he took it for granted that all persons interested in heat would have been familiar enough with Boerhaave's book to have made it unnecessary to refer to it. If this be so, then it dramatizes our general lack of familiarity today with what would have been considered the common knowledge of physical scientists of the eighteenth century.

[10] See Florian Cajori, *A history of physics in its elementary branches, including the evolution of physical laboratories* (New York: Macmillan Company, 1929), pp. 178 ff.

[11] E.g., Cajori (reference 10 above), p. 184.

A Note Concerning Diderot and Franklin†

By I. Bernard Cohen

PROFESSOR DIECKMANN's illuminating comparison of the two versions of Diderot's *Interprétation de la nature* calls attention anew to the problem of the relations between Diderot and Franklin. It will be recalled that in this book, Diderot went out of his way to praise Franklin's work — praise all the more noteworthy in that most men of science were cited by Diderot for criticism. Section XLI begins with the oft-quoted remarks about the veil of nature and the strictures about not doubling that veil with one of mystery: ". . . n'est-ce pas assez des difficultés de l'art?" Then there appears a laudation of Franklin and the chemists: "Ouvrez l'ouvrage de Franklin; feuilletez les livres des chimistes, et vous verrez combien l'art expérimental exige des vues, d'imagination, de sagacité, de ressources: lizes-les attentivement, parce que s'il est possible d'apprendre en combien de manières une expérience se retourne, c'est là que vous l'apprendrez."

This passage, identical in the 1753 and 1754 editions, was written at the time of the first outburst of Franklin's popularity in France. Franklin's book on electricity had appeared in a French translation made by Dalibard in 1752 at the request of Buffon.[1] The many ingenious experiments described by Franklin were repeated by many Frenchmen; Buffon and Dalibard, aided by Delor, performed the new Philadelphia experiments at St. Germain in the presence of Louis XV and his court. The great experiment proposed by Franklin was to test the possible electrification of clouds and so prove the electrical character of the lightning discharge. This sentry-box experiment was undertaken by the three experimenters, and success was first obtained by Dalibard's assistant on 10 May 1752 and reported by Dalibard to the Académie des Sciences eight days later. The king himself sent "his thanks and compliments . . . to Mr. Franklin of Pennsylvania."[2] So great was the general interest that in 1753 Dalibard brought out an enlarged version of Franklin's book, containing the report on the lightning experiments. The attacks by Abbé Nollet called fur-

† This note has been written to supplement the article of Herbert Dieckmann: The first edition of Diderot's *Pensées sur l'interprétation de la nature*, *Isis*, 1955, *46*: 251–267.

[1] For information concerning Buffon's part in the preparation of the French version of Franklin's book, see I. B. Cohen: "The two hundredth anniversary of Benjamin Franklin's two lightning experiments and the introduction of the lightning rod," *Proc. Amer. Philos. Soc.*, 1952, *96*: 338, n. 15. The editions are described in *Benjamin Franklin's experiments: a new edition of Franklin's "Experiments and observations on electricity,"* ed. by I. B. Cohen (Cambridge: Harvard University Press, 1940), intro., pt. 4.

[2] Further information concerning these events may be found in the two works cited in footnote 1 above. A résumé of the stages by which Franklin's book "produisit la plus grande sensation dans toute l'Europe sçavante," may be found in *Observations sur la physique . . .*, ed. by Rozier, vol. 2 (Sept. 1773), p. 204.

ther attention to Franklin's work. We may easily see, therefore, why Diderot was interested in Franklin's discoveries in 1752, 1753, and 1754. Electricity was a major subject of investigation at the mid-century; and to it Diderot devoted a number of pages in his *Interprétation de la nature*. But before examining the alterations made by Diderot in his account of electricity, we may inquire into the history of the delations between him and Franklin. Diderot's name does not appear in the published or unpublished letters of Franklin, nor does Franklin's name appear anywhere in the correspondence of Diderot. There is, furthermore, no record of their ever having met, despite Franklin's residence in France. The lack of personal acquaintance between them has seemed so curious that an imaginary meeting has been invented.[3]

In 1773, when Diderot went to Russia at the invitation of Catherine the Great, Franklin was elected an *associé étranger* of the Académie des Sciences and a much improved edition of his scientific and miscellaneous "philosophical" writings was published in Paris, prepared by Barbeu Dubourg.[4] A review of the new edition in Rozier's *Observations sur la physique . . .* called attention to the fact that the theory "de ce sçavant et laborieux Anglois est suffisamment connue en France" so that the reviewer could rest content with presenting only some of the new additions not to be found in the earlier French versions by Dalibard.

Yet it is curious to note that when Diderot drew up his "Plan d'une université pour le gouvernement de Russie," written in 1775–1776, in which he listed the great books of experiment ("physique expérimentale"), he did not include Franklin's book on electricity on which he had lavished so much praise two decades earlier.[5]

While Franklin was in France as the representative of America during the American Revolution, Diderot included a reference to him in "La pièce et la prologue." Originally, the lines read:

Monsieur Poultier

Je n'en doute pas, vous êtes excellent quand vous avez tort. Mais ces Insurgents nous tracassent, et il faut que j'aille passer la soirée à Passy.

Monsieur Hardouin

Avec Franklin? (M. Poultier fait signe de tête.) Quel homme est-ce?

Monsieur Poultier

Un acuto quackero.

In the Fonds Vandeul, there is a copy with corrections in Diderot's hand. Monsieur Poultier's speech now ends, "et il faut que j'aille . . . ." In Monsieur Hardouin's reply, the opening words "Avec Franklin?" are replaced by "A Passy," and Monsieur Poultier's rejoinder is expanded to read, "Comme on l'a dit: un acuto quackero." In the version printed by Assézat, Monsieur

[3] This may be found in Willis Steell: *Benjamin Franklin at Paris, 1776–1785* (New York: Minton, Balch & Company, 1928), pp. 30–34, 147, as taken from "a French *revue*."

[4] See *Benjamin Franklin's experiments* (reference 2, above), intro., pt. iv.

[5] "Plan d'une université pour le gouvernement de Russie ou d'une éducation publique dans toutes les sciences (en grande partie inédit)," *Oeuvres complètes de Diderot*, ed. J. Assézat (Paris: Garnier Frères, 1875), tome troisième, pp. 409–534.

Hardouin's speech now begins with the words "Voir leur patriarche" in place of "A Passy," and the words "un acuto quackero" are in italics.[6]

We know that Franklin owned a copy of the *Encyclopédie* and very likely may have read the flattering remarks about his work in such articles as *Foudre, Coup foudroyant, Conducteur de foudre,* which not only praised his research in electricity but credited him with having founded a new *art*: lightning protection. And no doubt he read such articles as *Feu électrique* and *Électricité,* in which much of his work was summarized.[7] Yet we do not possess a single scrap of information concerning Franklin's reaction to these articles, nor proof that he actually did read them. Even more vexing is the question of whether Franklin ever read Diderot's laudation in the *Interprétation de la nature.* The only reference to this book in Franklin's correspondence or his writings occurs in an unsigned note in the unpublished letters in the American Philosophical Society Library, attributed to Le Veillard.[8]

Louis Le Veillard, Franklin's neighbor at Passy, was a fellow member of the Loge des Neuf Soeurs in Paris.[9] When Franklin left Paris for home at the end of the American Revolution, Le Veillard accompanied him to Le Havre and went across the Channel with him to Southampton. In 1786, the American Philosophical Society elected Le Veillard to membership along with other French neighbors and friends of Franklin's. Finally, it will be recalled that it was to Le Veillard that Franklin sent a MS copy of his "Autobiography" in the hope of getting his "friendly, candid opinion of the parts you would advise me to correct or expunge."[10] The note in the American Philosophical Society Library, attributed to Le Veillard, is undated; it begins with the following reference to Diderot: "Dear Sir [:] I send you The Book of M. Diderot called *L'interprétation de la nature,* where you will find Those ridiculous Explanations of The causes of *the Aurora Borealis,* pag. 79, 80, 83." Very likely the occasion of this note was a paper read by Franklin on the cause of aurorae. Franklin completed his essay on the aurora borealis by December 1778, and, according to Carl Van Doren, it "seems to have been read at a meeting of the

[6] Diderot's manuscript emendations were communicated to me by Professor Dieckmann; see his *Inventaire du fonds Vandeul et inédits de Diderot* (Genève: Librairie Droz; Lille: Librairie Giard, 1951); *Oeuvres complètes de Diderot,* tome huitième, pp. 69–133.

[7] Yet Franklin would have raised at least one eyebrow had he read, *s. v.* "Tonnerre," the following statement: "On peut rompre & détourner le tonnerre par le son de plusieurs grosses cloches, ou en tirant le canon; par-là on excite dans l'air une grande agitation qui disperse les parties de la foudre; mais il faut bien se garder de sonner lorsque le nuage est précisément au-dessus de la tête, car alors le nuage en se fendant peut laisser tomber la foudre." In Tome quatrième of the *Supplément à l'Encyclopédie . . .* published in Amsterdam in 1777, *s.v.* "Tonnerre," Guyton de Monveau made only a passing reference to the practice of ringing bells during thunderstorms, devoting most of the space assigned to him to details of lightning-rod construction; the article began with a reference to the Barbeu Dubourg translation of the writings of "le célèbre Franklin." In Tome troisième of the *Supplément,* about one-half of the article on "Foudre" is a résumé of Franklin's writings, with a long quotation, and the remainder is based on the work of Beccaria. Further information about the lightning experiments of Franklin, Beccaria, and others may be found, *s.v.* "Cerf-volant."

[8] It is described in I. Minis Hays: *Calendar of the Papers of Benjamin Franklin in the Library of the American Philosophical Society* (5 volumes, Philadelphia: American Philosophical Society, 1908).

[9] See Nicholas Adolf Hans: "Unesco of the 18th century, La loge des neuf soeurs and its venerable master, Benjamin Franklin." *Proc. Amer. Philos. Soc.,* 1953, *97*: 513–524.

[10] See Carl Van Doren: *Benjamin Franklin* (New York: Viking Press, 1938), pp. 656, 659, 723–726, 766–767.

Academy of Sciences after Easter of the following year." [11] It was published in both French and English.[12]

In Le Veillard's letter, no mention was made of Diderot's praise of Franklin, which occurs in *L'interprétation de la nature* several pages before the references to the *Aurora Borealis* to which Le Veillard referred Franklin. We do not have available Franklin's comments on *L'interprétation de la nature*, and perhaps they were given verbally to Le Veillard. Hence we not even know if Franklin ever read any of the book. In the 1754 edition of this book Diderot had asserted that ". . . les Aurores Boréales . . . ne sont probablement que des courants de matiere électrique." But Professor Dieckmann has shown us that the first version (1753) had an additional phrase: ". . . comme on ne manquera pas de s'en assurer avec le cerf-volant du Philosophe de Philadelphie." The experiment of the electrical kite had shown that clouds are electrified and that the lightning discharge is, therefore, an electrical phenomenon. Diderot's assumption that the kite experiment was related to auroral phenomenon was unwarranted, and he acted wisely in suppressing it. Another suppression was XXXVII, 6e REVERIE. This began: "Toutes les expériences qu'on a faites jusqu'à présent, l'ont été hors du globe électrique." Actually Diderot was in error, because Hauksbee had performed a number of experiments inside the rotating globe. Presumably this work had been brought to his attention only after 1753, although it was mentioned in the historical account of electricity prefaced to the 1752 French edition of Franklin's book on electricity prepared by Dalibard.

The long section immediately following the references to the aurora borealis in the 1753 edition of *L'interprétation de la nature*, and suppressed in the 1754 edition, depend upon an experiment with a covered globe and Diderot said that if it were to fail, the system he proposed would have to be abandoned. It would be interesting to know whether he had actually performed the experiment, had had it performed for him, or simply had come to the conclusion that the whole idea was worthless. There is no information available. Diderot was certainly wise in suppressing the beginning of XXXV, 4^e^ REVERIE, dealing with the possible use of electricity in curing male sterility. His substitution of a conjectured similarity between fire and the electric fluid was an ingenious improvement: his well-known interest in chemistry was displayed in his supposition that the electric fluid might augment the weight of metals in the same way that the calcination of lead by fire increases its weight.

Many points of similarity between Bacon and Diderot have been remarked.[13] Like Bacon, Diderot largely wrote about experiment rather than performing useful experiments. But Diderot was more interested in the latest examples of genuine experiment than Bacon was. Bacon never appreciated the great work

[11] *Ibid.*, p. 658.

[12] "Des suppositions & des conjectures sur la cause des aurores boréales de M. Francklin" was published in *Observations sur la physique* . . . . 1779, *13*: 409–412. A footnote reads as follows: "Ce mémoire a été lu le 14 avril, par M. Le Roy, à la séance publique de l'Académie Royale des Sciences. Les sujets importans dont ce Ministre Plenipotentiaire est occupé, prouvent que le génie de sciences n'exclut pas celui des affaires."

[13] See Herbert Dieckmann: "Influence of Francis Bacon on Diderot's *Interprétation de la nature*." *Romanic Review*, 1943, *34*: 303–330.

of his countryman Gilbert on magnetism and electricity,[14] just as he did not properly value the great discoveries of Galileo, Kepler, or Harvey.[15] But Diderot was keenly aware of the currents of research in the 1750's, and his discussion of electricity in the 1753 edition of his *Interprétation de la nature*, tempered and wisely modified in the revised 1754 edition, shows him to have been sensitive to the most recent discoveries in experimental physical science.

[14] See Duane H. D. Roller: "Did Bacon know Gilbert's *De Magnete*?" *Isis*, 1953, *44*: 10–13.

[15] See Jean Pelseneer: "Gilbert, Bacon, Galilée, Képler, Harvey, et Descartes: leurs relations." *Isis*, 1932; *17*: 171–208.

Benjamin Franklin and Jonathan Edwards on Lightning and Earthquakes‡

BY ALFRED OWEN ALDRIDGE *

DR I. Bernard Cohen has discovered that an essay on the cause of earthquakes printed by Benjamin Franklin in the *Pennsylvania Gazette* 8–15 December and 15–22 December 1737, numbers 470–71, contains an explanation of lightning essentially the same as one written by Jonathan Edwards in his diary.[1] Because of the remarkable similarity of these two accounts, Cohen raised the question as to whether they derived from the same source. The essay in the *Pennsylvania Gazette* has been considered one of Franklin's earliest scientific writings, but it may now be demonstrated that the essay was not by Franklin at all and that Edwards probably derived his ideas on lightning from the same English work from which Franklin extracted his discussion.

In the *Pennsylvania Gazette* we read that:

> 1. The earth itself may sometimes be the cause of its own shaking. . . . 2. The subterraneous waters may occasion earthquakes by their overflowing. . . . 3. The air may be the cause of earthquakes. . . . Lastly, fire is a principal cause of earthquakes. . . .
>
> . . . Dr. Lister is of opinion, that the material cause of thunder, lightning, and earthquakes, is one and the same, viz. the inflammable breath of the pyrites, which is a substantial sulphur, and takes fire in itself.

Edwards writes in his diary:

> Lightning seems to be this: An almost infinitely fine, combustible matter, that floats in the air, that takes fire by a sudden and mighty fermentation, that is some way promoted by the cool and moisture, and perhaps attraction, of the clouds. By this sudden agitation, this fine, floating matter, is driven forth with a mighty force one way or other, which ever way it is directed, by the circumstances and temperature of the circumjacent air; for cold and heat, density and rarity, moisture and dryness, has almost an infinitely strong influence upon the fine particles of matter. This fluid matter, thus projected, still fermenting to the same degree, divides the air as it goes, and every moment receives a new impulse by the continued fermentation; and as its motion received its direction, at first, from the different temperature of the air, on different sides, so its direction is changed, according to the temperature of the air it meets with, which renders the path of the lightning so crooked. The parts are so fine, and are so vehemently urged on, that they instantaneously make their way into the pores of earthly bodies, still burning with a prodigious heat, and so instantly rarifying the rarifiable parts. Sometimes these bodies are somewhat bruised; which is chiefly by the beating of the air that is, with great violence, driven every way by the inflamed matter.[2]

An earthquake occurred in the Middle Atlantic states on 7 December, 1737. In the subsequent issue of the *Pennsylvania Gazette,* Franklin introduced his feature article with the comment:

> The late earthquake felt here and probably in all the neighbouring provinces, have made many people desirous to know what may be the natural cause of such violent concussions; we shall endeavour to gratify their curiosity by giving them the various opinions of the learned on that head.

At this point Franklin begins a word by word transcript from the article "Earthquakes" in Ephraim Chambers' *Cyclopaedia: or, an Universal Dictionary of Arts and*

‡ This note was written with the assistance of a grant from the Penrose Fund of the American Philosophical Society.

* University of Maryland, College Park, Md.

[1] *Benjamin Franklin's Experiments* (Cambridge, Massachusetts, 1941), pp. 109–111.

[2] Quoted by Cohen from S. E. Dwight, *The Life of President Edwards* (New York, 1830), Appendix, p. 743.

Sciences . . . (London, 1728). Franklin began copying at a point near the middle of Chambers' account, perhaps because he intended at first to devote only one issue to the subject. Then he may have decided that current interest in earthquakes demanded that Chambers' article be printed in entirety. At any rate, in the subsequent issue he printed the rest of the article, adding at the end the material that he had previously omitted.

We know that Edwards also read Chambers' *Cyclopaedia* and that he used it, moreover, in his literary work. His initial knowledge of the English philosopher Francis Hutcheson, for example, came to him from the article "Beauty" in Chambers.[3] It is quite likely, therefore, that his basic scientific knowledge derived from the same source. Judged only from the two men's interest in lightning, Edwards at this time had reached a more original and independent method of scientific inquiry than had Franklin, for Edwards does not limit his speculation to the theories in Chambers. The *Pennsylvania Gazette* supports other biographical evidence that Franklin had little leisure to devote to scientific speculation or experiment until 1743, when his circular letter on exchanging scientific information led to the formation of the American Philosophical Society.

Five weeks after Franklin printed the article on earthquakes, it was reprinted by his fellow publisher William Parks in the *Virginia Gazette* with an introduction very much like Franklin's.

> We having in our last given an Account of the Earthquake felt in several Places to the Northward, and many People being desirous to know what may be the natural Cause of such violent Concussions, we shall endeavour to gratify their Curiosity, by giving them the various Opinions of the Learned on that Head.[4]

We know that Parks took his material from Franklin instead of directly from Chambers because he followed Franklin's inverted order instead of the order of the original. He also reprinted from Franklin's news columns an account of the earthquake in Philadelphia, where it had been less violent than in New York. This is Franklin's only original writing on earthquakes at this time. For three or four evenings successively after the earthquake, "an unusual Redness appeared in the Western Sky and southwards, continuing about an Hour after Sunset, gradually declining." The quake was not felt in Annapolis; but at Newcastle, Delaware, and at Conestogoe, one hundred miles west of Philadelphia, the tremors were as violent as in Philadelphia. From a source in New York, Franklin reported that "the first Sense we had of it was like a strong Gale of Wind, which encreased 'til it began to resemble the Noise of Coaches swiftly driven."

Chambers had been extremely useful to Franklin on subjects other than earthquakes. When the *Pennsylvania Gazette* had been published by Samuel Keimer under the title of *The Universal Instructor in all Arts and Sciences: and Pennsylvania Gazette*, at least half of its contents consisted of extracts from Chambers. In forty issues Keimer had published everything in Chambers from A to AIR. When Franklin became editor, 25 September 1729, he announced that the dictionary from Chambers would be discontinued; then despite this disavowal, he later reprinted Chambers' articles on hemp (No. 42), on inoculation for smallpox (No. 80), and on Free-Masonry (No. 130).

One other essay in the *Pennsylvania Gazette* which is sometimes taken as evidence of Franklin's early scientific pursuits is "Of the late wonderful Discoveries, and Improvements of Arts and Sciences," which appeared 14 October 1736, No. 409. This

---

[3] Thomas H. Johnson, "Jonathan Edwards' Background of Reading," *Publications of the Colonial Society of Massachusetts*, XXVIII (December, 1931), 204.

[4] 20–27 January, 27 January–3 February, 1737, Nos. 78–79. Because before 1753 the new year did not begin until 25 March, January 1737 (or January 1937/8), was obviously the month following December 1737.

essay may show that Franklin had an interest in science, but it provides no more evidence than does the essay on earthquakes of Franklin's independent investigations, for like that essay it is merely an extract from another source. It appeared originally in *The Prompter*, a London literary periodical, 11 June 1736, No. 167. Before Franklin it had been reprinted moreover in the *London Magazine*, June, 1736, and in the *New England Weekly Journal*, 21 September 1736, No. 494; after Franklin it was printed by William Parks in the *Virginia Gazette*, 7–14 January 1736, No. 24.

Benjamin Franklin, Thomas Malthus and the United States Census

By Conway Zirkle *

THE object of this note is to call attention to (1) an inspired guess, (2) a remarkable prediction and (3) a still more remarkable coincidence. The field in which the guess, the prediction and the coincidence occurred was that of population growth, and the man who was responsible for it all was Benjamin Franklin. As nearly as we can judge, he made the guess with almost no data to guide him, he based his prediction on his guess, and his guess, applied to our census figures, produced the coincidence. The whole affair is rather odd because his prognostication of our population growth corresponds so closely with the actual growth shown by the census that all the calculations can be made without resort to logarithmic tables.

The fact that all species of animals and plants have the potentiality of increasing their numbers without limit was recognized in the seventeenth century and, in the eighteenth, it became a valuable generalization in biological thought. Also the fact was well known that a stable balance exists between the species in nature, and that the number of individuals in each species remains practically constant except, of course, for seasonal variations or when the balance of nature is upset by some catastrophe. This balance is the result of conflicting forces, which were easily identified. In 1677, Sir Matthew Hale described the over-production of offspring and the "correcting" activities in nature, which restores the balance, in his well known book, *The Primitive Origination of Mankind* (London, 1677, page 211). During the next century, the over-production of young, and the consequent population pressure which limits their numbers, were described by Buffon (1751), Franklin (1755), Bonnet (1764), Monboddo (1773), Herder (1784) and Smellie (1790).[1] The fact was also recorded that mankind itself, as well as the animals and plants, had the potentiality of an unlimited increase in numbers, and that, if their numbers were not checked, they would increase indefinitely.

By far the most famous of all works on population increase was the *Essay on the Principles of Population* written by Thomas Malthus and published for the first time in 1798. Malthus pointed out that any population *growing unchecked* would increase in geometrical progression. He was right, of course, but he was not the first to make this observation. It had been implied earlier by Benjamin Franklin in 1755, and shown specifically to be the case by Robert

* University of Pennsylvania.

[1] The relevant passages have been quoted elsewhere. See Conway Zirkle: "Natural Selection before the *Origin of Species*," *Proc. Am. Phil. Soc.*, 1941, *84*: 71–123.

Wallace in an often overlooked work, *A Dissertation on the Numbers of Mankind* (Edinburgh, 1753). This geometrical increase in *unchecked* human population is today an accepted truism and is implied in our method of describing population growth. Every population that is growing at a constant rate — no matter what that rate may be — is increasing in a geometrical series (exponentially). Bacteria inoculated into a fresh broth culture, rabbits released in Australia, and human beings in a newly discovered and relatively empty continent all increase exponentially for a time.

Malthus was aware of the fact that no population can actually grow in this manner except for a short time and in a limited region, and he concluded naturally that population growth is always checked and that the normal tendency of population is to press upon the checks and to increase whenever any check is reduced or removed. But to measure the checks on population he had to have a control, a population that was growing unchecked — under conditions where the limiting factor of population pressure did not operate.

There was but one country at the time that met the requirements for Malthus' control. It was the United States of America. But here Malthus was faced with a real difficulty. When he published the first edition of his famous *Essay*, the United States had taken but one census, and a single census obviously could not give him the rate of population growth. By the time his second edition was published in 1803, the United States had taken its second census, and the rate of increase for the ten years was 35.1% (3.05% per year). This rate indicated that the population was doubling every 23 years, 15 days. A single decade's increase, however, was a rather shaky foundation for an estimate of a growth rate, and Malthus accepted a more conservative guess made by Benjamin Franklin. (We really have no evidence that Malthus ever based any calculations on the two censuses of the United States.)

The relevant passage from the *Essay on the Principles of Population* is worth quoting, for its meaning has generally been missed, and those who should know better have often misrepresented it. It occurs in the very first chapter of the second and all subsequent editions and is as follows:

> The cause to which I allude is the constant tendency in all animated life to increase beyond the nourishment prepared for it.
>
> It is observed by Dr. Franklin, that there is no bound to the prolific nature of plants or animals but what is made by their crowding and interfering with each other's means of subsistence. Were the face of the earth, he says, vacant of other plants, it might be gradually sowed and overspread with one kind only, as, for instance, with fennel; and were it empty of other inhabitants, it might in a few ages be replenished from one nation only, as, for instance, with Englishmen. This is incontrovertibly true. Throughout the animal and vegetable kingdoms Nature has scattered the seeds of life abroad with the most profuse and liberal hand; but has been comparatively sparing in the room and the nourishment necessary to rear them. The germs of existence contained in this earth, if they could freely develop themselves, would fill millions of worlds in the course of a few thousand years. Necessity, that imperious, all-pervading law of nature, restrains them within the prescribed bounds. The race of plants and the race of animals shrink under this great restrictive law; and man cannot by any efforts of reason escape from it.
>
> In plants and irrational animals, the view of the subject is simple. They are

all impelled by a powerful instinct to the increase of their species, and this instinct is interrupted by no doubts about providing for their offspring. Whenever, therefore, there is liberty, the power of increase is exerted, and the superabundant effects are repressed afterwards by want of room and nourishment.

The effects of this check on man are more complicated. Impelled to the increase of his species by an equally powerful instinct, reason interrupts his career, and asks him whether he may not bring beings into the world for whom he cannot provide the means of support. If he attends to this natural suggestion, the restriction too frequently produces vice. If he hears it not, the human race will be constantly endeavoring to increase beyond the means of subsistence. But as, by that law of our nature which makes food necessary to the life of man, population can never actually increase beyond the lowest nourishment capable of supporting it, a strong check on population, from the difficulty of acquiring food, must be constantly in operation. The difficulty must fall somewhere, and must necessarily be felt in some or other of the various forms of misery, or the fear of misery, by a large portion of mankind.[2]

Malthus next attacks the problem of finding out just how fast a human population would increase without checks and here he uses Franklin's guess.

That population has this constant tendency to increase beyond the means of subsistence, and that it is kept to its necessary level by these causes, will sufficiently appear from a review of the different states of society in which man has existed. But, before we proceed to this review, the subject will perhaps be seen in a clearer light, if we endeavour to ascertain what would be the natural increase of population, if left to exert itself with perfect freedom; and what might be expected to be the rate of increase of productions of the earth, under the most favorable circumstances of human industry. . . .

In the northern states of America, where means of subsistence have been more ample, the manners of the people more pure, and the checks to early marriages fewer, than in any of the modern states of Europe, the population has been found to double itself, for above a century and a half successively, in less than twenty-five years. Yet, even during these periods, in some of the towns, the deaths exceeded the births, a circumstance which clearly proves that, in those parts of the country which supplied the deficiency, the increase must have been much more rapid than the general average.

This brings us to a remarkably prescient paper written by Franklin in 1751 and published in Boston in 1755. It is entitled *Observations Concerning the Increase of Mankind and the Peopling of Countries.* Here Franklin estimates that the population of the American Colonies doubles every twenty-five years and that within a hundred years more Englishmen will be west of the Atlantic Ocean than east of it. He also made the sound but still not understood observa-

[2] The fact that Malthus followed Franklin in applying to animals and plants the tendency of breeding beyond the capacity of nature to feed them is important on several counts. (1) It gave Charles Darwin the clue as to how nature could select some of those born and eliminate others. It was this clue that Darwin got from Malthus twenty years before the *Origin of Species* was published that led Darwin to his concept of natural selection. Incidentally, Alfred Russell Wallace, who discovered natural selection independently, was also led to his discovery by reading Malthus in 1858. (2) It is also a passage that Karl Marx misrepresented several times. In a letter to Engels dated 18 June 1862 Marx wrote, "I am amused by the statement of Darwin, whom I am reading now, that he applies the 'Malthusian' theory to plants and animals also, whereas the whole point of Mr. Malthus lies in the fact that he does *not* apply his theories to plants and animals but *only* to man — with geometrical progression — as opposed to plants and animals." Marx repeats this misstatement in his *Theories of Surplus Values.* It is probable that Marx never read Malthus, but it is possible that (a) either Marx read Malthus without understanding him, or that (b) Marx read and understood Malthus but knowingly misrepresented him.

tion that emigration does not reduce the population of a country and that immigration does not increase it. (The rate of population increase in the United States was *greatest* before the large, late nineteenth-century immigrations.) Franklin was trying at the time to convince the British of the importance of the American Colonies and that the British exploitation of the Colonies was stupid, as, of course, it was. The relevant portion of Franklin's paper is in Sections 22 and 23:

> 22. There is, in short, no bound to the prolific nature of plants or animals, but what is made by their crowding and interfering with each other's means of subsistence. Was the face of the earth vacant of other plants, it might gradually be sowed and overspread with one kind only, as, for instance, with fennel; and, were it empty of other inhabitants, it might in a few ages be replenished from one nation only, as, for instance, with Englishmen. Thus, there are supposed to be now upwards of one million English souls in North America (though it is thought scarce eighty thousand has been brought over sea), and yet perhaps there is not one the fewer in Britain, but rather many more, on account of the employment the colonies afford to manufacturers at home. The million doubling, suppose but once in twenty-five years will, in another century, be more than the people of England, and the greater number of Englishmen will be on this side of the water. What an accession of power to the British empire by sea as well as land! What increase of trade and navigation! What numbers of ships and seamen! We have been here but little more than one hundred years, and yet the force of our privateers in the late war, united, was greater, both in men and guns, than that of the whole British navy in Queen Elizabeth's time. How important an affair then to Britain is the present treaty for settling the bounds between her colonies and the French, and how careful should she be to secure room enough, since on the room depends so much the increase of her people.
>
> 23. In fine, a nation well regulated is like a polypus; take away a limb, its place is soon supplied; cut it in two, and each deficient part shall speedily grow out of the part remaining. Thus, if you have room and subsistence enough, as you may, by dividing, make ten polypuses out of one, you may of one make ten nations, equally populous and powerful; or rather increase a nation ten fold in numbers and strength.

We can now use the United States Census to measure the accuracy of Franklin's estimate of the rate of population growth. The first census taken in 1790 showed a population of 3,929,214, and the second census of 1800 showed a population of 5,308,483, the rate of increase in the decade being 35.1%. This was approximately the rate of increase until 1860, the year of the last census before the Civil War. The increases for these seven decades were, respectively, 35.1%, 36.4%, 33.1%, 33.%, 32.7%, 35.9% and 35.6%, the average increase being 34.6%. During these seventy years the population was growing at an almost constant rate and consequently it was increasing in an exponential manner. Subsequent censuses showed a marked and almost constant decrease in rate of growth. (Some later censuses are known to have been very inaccurate and hence there are some unreal fluctuations shown in the reported growth rates.)

We can now raise the question, how long did it take our population to double during the period of its exponential growth? Starting with the first census, the population was 3,929,214. The first doubling would come to 7,858,428; the second doubling would bring the number to 15,716,856; and the third to 31,433,712, and this brings us to the first coincidence. The last census of this seventy-year growth period, the census of 1860, showed our population to be 31,443,321! Eight (2^3) times the census of 1790 gives a figure within three-hundredths of one per cent of the census of 1860! Our population had doubled then just three times in seventy years. If we divide the seventy years by three, the answer is twenty-three years and four months for the doubling period. Franklin's guess of twenty-five years was not bad.

At first glance Franklin does not seem to have been so fortunate in his 1751 prediction that more Englishmen would be in America than in England in one hundred years' time. The 1851 census of Great Britain and Ireland showed a population of 27,513,551, while the United States in 1850 had a population of 23,191,876, and not all of them were English. If we restrict the two populations to the English, however, as Benjamin Franklin did, we find that in 1851 England and Wales had only 17,900,000 people. The United States census of 1850 included 3,638,808 Negroes and this left a white population of 19,553,068. The excess of the United States white over the English and Welsh was one million six hundred thousand, and this was doubtless enough to take care of the non-English whites — the "Palatine boors," whom Franklin disliked, and the first migrations of the Irish which followed the famine of 1846. Again Franklin seems to have been right. There were more English in America than in England in 1850.

We come finally to a truly remarkable coincidence. It can be found in a calculation that Franklin himself could not make because he died too soon — he died in 1790, the year the first census was taken. But let us suppose that a Franklinian or a group of Franklinians wished to look a hundred years into the future and to learn what our population would be in 1890. Suppose they applied Franklin's estimate of our population doubling every twenty-five years to the figures given by the first census and had used their calculations to foretell the population at the next century mark. Their estimates of the future might well startle us. The 1790 population of 3,929,214 would — according to formula — double just four times in the 100 years and thus the Franklinians would multiply the 1790 figures by 16. The answer would have been 62,867,424. The census count showed that our population in 1890 was 62,947,714. The calculated and the counted population differ from each other by just thirteen hundredths of one per cent (.13%).

We will, of course, dismiss this congruence as a coincidence, for we know that the calculation contained two errors which just *happened* to cancel out each other exactly. During the first seventy years of our national existence, when our population was growing at an almost constant rate, it doubled three times, or five years before it should have according to Franklin's formula. During the next thirty years its growth slowed down, and it required five more years to double the fourth time than it should have according to the formula. The precision of the end result is only an accident. (But is it?) We must remember that Franklin was discovery-prone. Major prophets have the exasperating habit of being right.

Franklin's Introduction to Electricity

By N. H. de V. Heathcote *

IN a paper published some years ago, Professor I. Bernard Cohen [1] discussed the identity of the "Dr. Spence" mentioned by Franklin in his *Autobiography*. He established beyond all reasonable doubt that "Dr. Spence" was in fact a certain Dr. Adam Spencer who offered a course of lectures in experimental philosophy in Boston in 1743 and gave several courses in Philadelphia in 1744. Franklin, who was at that time postmaster in Philadelphia, met Spencer in Boston in 1743 and acted as his agent in Philadelphia in 1744; [2] Professor Cohen considers it probable that he attended Spencer's lectures.[3]

Closely connected with the question of "Dr. Spence's" identity is the double question: When and through whom did Franklin first become acquainted with the subject of electricity? Was it through Spencer in 1743 or through Peter Collinson of the Royal Society of London in 1746? Professor Cohen, after a most careful analysis of the evidence, pronounces in favour of Spencer in 1743.[4] There are grounds, however, for believing that this conclusion is only partially correct. As Professor Cohen's article was published some time ago and in another periodical, it will be as well to set out here the evidence on which he bases his conclusion.

In the *Boston Evening Post* of 30 May 1743 Spencer advertised a "Course of Experimental Philosophy" to be given in Boston. Apparently the requisite number of subscribers did not come forward, and the lectures were not given. Spencer next appears in Philadelphia, where he was more successful, for a notice appeared in Franklin's *Pennsylvania Gazette* of 26 April 1744 to the effect that more had subscribed to Dr. Spencer's first course than could be accommodated, and that a second course would begin in May.[5]

In his *Autobiography* Franklin says [6] that in 1746 he met in Boston "a Dr. Spence, who . . . show'd me some electric experiments. They were imperfectly perform'd, as he was not very expert; but, being on a subject quite new to me, they equally surpris'd and pleas'd me," and adds that soon after his return to Philadelphia the Library Company of Philadelphia received a present of a glass tube and instructions for its use from Peter Collinson.[7]

In a letter to Peter Collinson's son Michael, written shortly after the father's death in 1768, Franklin says he received the glass tube and instructions for its

* Department of the History and Philosophy of Science, University College, London.

[1] Cohen, I. Bernard: Benjamin Franklin and the Mysterious "Dr. Spence." *Journal of the Franklin Institute*, 1943, *235*: 1–25.

[2] *Ibid.*, p. 5.

[3] *Ibid.*, p. 17.

[4] *Ibid.*, p. 20.

[5] *Ibid.*, p. 4.

[6] *Ibid.*, p. 2.

[7] *Ibid.*, p. 2.

use from Collinson in 1745 and that "this was the first Notice I had of that curious Subject, which I afterwards prosecuted with some Diligence." [8]

Professor Cohen is undoubtedly correct in regarding the date 1746 in the *Autobiography* as an error, not necessarily on Franklin's part; that Franklin intended the date to be given as 1743 is evident from his "Draft Scheme" of the *Autobiography*, where he mentions going to Boston in 1743 and seeing "Dr. Spence." [9] With regard to the date 1745 in the letter to Michael Collinson, this, as Cohen suggests, was almost certainly an error on Franklin's part, owing to the fact that more than twenty years had elapsed since the event.

Franklin's first letter to Collinson is dated 28 March 1747. In it he says: "Your kind present of an electric tube, with directions for using it, has put several of us on making electrical experiments. . . . For my own part, I never was before engaged in any study that so totally engrossed my attention and my time as this has lately done; for what with making experiments when I can be alone, and repeating them to my Friends and Acquaintance, who, from the novelty of the thing, come continually in crowds to see them, I have, during some months past, had little leisure for any thing else." [10] This suggests some time in the second half of 1746 for the receipt of the present, in agreement with the date given in the passage from the *Autobiography* quoted above.

To summarize: Spencer was in Boston in 1743; Franklin says he saw him there in that year, if we accept the date given in the "Draft" rather than that given in the *Autobiography* itself; Spencer did not give his lectures in experimental philosophy in Boston, but did show Franklin some experiments (Franklin's statement in the *Autobiography*). Spencer was in Philadelphia in 1744 and gave at least two courses of lectures there; Franklin may have attended the lectures, though he does not mention having done so, and certainly acted as his agent. Collinson's gift reached the Philadelphia Library Company some time in 1746, and Franklin says in his letter to Michael Collinson written some twenty years later, that this was the first notice he had of electricity, a statement not inconsistent with the letter to Peter Collinson written within a few months of the receipt of the gift. Yet in the *Autobiography* he says that Spencer had shown him some "electric experiments" in Boston (in 1743).

Is it possible to reconcile these conflicting statements? It seems that it should be; for Franklin, though he might easily become confused as to dates, was hardly likely to forget what it was that started him on his brilliant career as an "electrician." It is, of course, not impossible that the statement in the letter to Michael Collinson was merely a "polite fiction," but this does not seem very probable. Again, the letter to Peter Collinson of 1747 does not explicitly state that the experiments with the tube were quite new to Franklin, merely that the gift and instructions led to a period of enthusiastic experimentation. But if Franklin had seen similar experiments in Boston three years earlier,

[8] *Ibid.*, p. 3.

[9] *Ibid.*, p. 20, quoting "Franklin's Draft Scheme of the Autobiography," published by A. H. Smyth in vol. 1 of *The Writings of Benjamin Franklin*, New York, 1905–07.

[10] *Benjamin Franklin's Experiments: A New Edition of Franklin's "Experiments and Observations on Electricity,"* edited, with a critical and historical introduction, by I. Bernard Cohen (Cambridge: Harvard University Press, 1941), p. 169.

why the lapse of time before the outburst of enthusiasm? True, it may have been owing to the lack of a suitable glass tube, but Spencer was in Philadelphia in 1744 and so was Franklin; yet there is no mention of any "electric experiments" in that year in Franklin's writings. Another point to be borne in mind is that on the three occasions when Franklin mentions the receipt of the glass tube, he also mentions the instructions as to its use; yet he presumably had already seen a tube actually in use, so why the emphasis on the instructions? After all, the experiments which could be made with the glass tube were very simple, so much so that a man of Franklin's calibre would hardly need to postpone making the experiments for himself until chance placed written instructions in his hands, as the three-year gap between 1743 and 1746 suggests he did. A more likely explanation seems to be that the experiments Spencer showed Franklin were of such a nature that they entirely failed to arouse Franklin's interest, and that it was only later that he realized their true significance. It is this possibility which I wish to examine here.

The experiments which Spencer showed Franklin privately would in all likelihood be those he was in the habit of performing in his lectures. We know what these were from two sets of notes, one certainly, the other probably, based on the lectures. These are quoted in full in Professor Cohen's article. The first is contained in the journal of William Black under the date 29 May 1744; it consists of a few lines only. Black says that Dr. A. Spencer gave a "Philosophical Lecture" on the eye and its diseases and then goes on to say:

> Next he proceeded to show that Fire is Diffus'd through all Space, and may be produced from all Bodies, Sparks of Fire Emitted from the Face and Hands of a Boy Suspended Horizontally, by only rubbing a Glass Tube at his Feet.[11]

The second source of information is some manuscript notes by John Smith, a Quaker interested in science. These notes, which cover both sides of two sheets of paper, are published for the first time in Cohen's article.[12] They are not labelled, but, since Smith was in Philadelphia in 1744, it is a reasonable surmise on Cohen's part that they were notes of Spencer's lectures. One page of the notes mentions the time taken by light to reach the earth from the sun; the eye and its diseases; heat produced by mixing various substances; that a bullet is warmed by friction with the air; that the strongest winds are always attended "with a hot Air"; that clouds of ice melting together in the air by their attrition cause lightning and "this heat" rarefies the air, thunder being caused by fresh air rushing in. He then goes on:

> Fire is diffused thro' all Space, Contained in, & may be produced from all bodies. A Boy was suspended Horizontally & the D[r]. rubbed a Glass Tube, a little distance from his feet w[ch]. made Sparks of fire fly from his face & hands.[13]

Both sets of notes make Spencer introduce the suspended-boy experiment at the conclusion of a lecture in which he had dealt with light, the eye, and

[11] Cohen, *loc. cit.*, p. 6, quoting *The Pennsylvania Magazine of History and Biography*, 1877, *1*: p. 246.

[12] *Ibid*, p. 10, note, where it is stated that the originals are in the Ridgway Library of the Library Company of Philadelphia, Smith MSS, vol. 5, pp. 254–255.

[13] *Ibid.*, p. 10.

heat (or fire); there is no reference here to any of the stock experiments (well known at that time, at least in Europe) on "electrical attraction." Both Black's and Smith's notes make it clear that Spencer in the latter part of the lecture was dealing with heat, while Smith's more detailed notes show that considerable emphasis was placed on the heat (fire) produced by friction, concluding with the sparks of fire from the boy when a rubbed tube was held near his feet.

That Spencer should treat this particular experiment as an instance of the universality of "fire" was quite in keeping with the ideas of the time. Increasing familiarity with the electric spark tended to foster the belief that there was a close connection between electricity and "fire," even that the two were in reality identical. Thus we read in s'Gravesande's *Mathematical Elements of Natural Philosophy*:

> There are several very remarkable Phenomena, which are ascrib'd to the Fire contain'd in Bodies; some of which ought to be mention'd here: Amongst which there are such as have a near Relation to Electricity; for which Reason we must also treat of the Phenomena of Electricity. . . . I shall only mention a few Experiments, made above thirty Years ago in England, to make manifest the Connexion between the Cause of Electricity and Fire, if these two ought really to be distinguish'd.[14] [And in the Preface he says:] Speaking of Fire, I have occasionally deliver'd a few Things about Electricity, that it might appear, that there is a Connexion between some Phaenomena of Fire and Electricity.[15]

Neither Adam's nor Smith's notes on the suspended boy experiment contain any suggestion that the experiment was of an electrical nature. There is, however, an explicit mention of electricity on another page of Smith's notes. At the head of the page is: "One of the D^{rs}. Experiments relating to Electrical Attraction was thus [*sic*]." The note then goes on:

> He took a Long Glass Tube, & Rubbed it Vehemently with his hand, & then held it pretty near several Pieces of Leaf brass or Gold, which put them into very brisk & Surprizing Motions. Some would leap toward the Tube, Sometimes adhere & fasten to it, settle on its Surface, and there remain Quiet: and sometimes be thrown off from it with a great force. And thus would they be alternately attracted and Repelled, for several times Successively. Sometimes, again, they would move slowly toward the tube, sometimes, would remain suspended between the tube and the Block they were first laid on; and sometimes slide along in the direction of the side of the tube, without touching it.[16]

The remaining notes on this page deal with arguments for the existence of a vacuum, expansion of liquids and solids, and so on.

If Smith's note is an accurate description of Spencer's experiment, then the audience saw an unmistakably "electric" experiment, and so, no doubt, did Franklin, either then or in Boston the year before. But are Smith's notes reliable on this point? If we turn to the paper published in 1706, in which

[14] *Mathematical Elements of Natural Philosophy, confirm'd by experiments: Or, an Introduction to Sir Isaac Newton's Philosophy.* Written in Latin by the late W. James s'Gravesande, LL.D. Translated into English by the late J. T. Desaguliers. Sixth Edition. 2 vols. (London, 1747), *2*: p. 72. The translation is from the 3rd Latin Edition.

[15] *Ibid.*, vol. 1, p. xxxi.

[16] Cohen, *loc. cit.*, p. 8.

Hauksbee first described the glass tube and leaf brass experiment, we read the following:

> I took a Hollow Tube of fine Flint Glass, about an Inch Diameter and 30 in Length, which having rubb'd pretty smartly with Paper in my Hand, till it had acquir'd some degree of Heat; it was then held towards some pieces of Leaf Brass, which so soon as its *Effluvium* had reacht, became suddenly in Motion, flying towards the Tube, even at 9 or 10 Inches distance; and it seem'd that the hotter the Tube was made by Rubbing, the farther it would Attract, but that it would do so to any Degree of Heat, I dare not determine. And what further observable was, That sometimes the Bodies Attracted would adhere to the Tube, and there remain quiet: Sometimes would be thrown violently from it to good Distances: Sometimes in their Motions towards, and sometimes even touching it, they would suddenly be Repell'd back to the distance of 4 or 5 Inches, repeating the same several times with great Velocity in a very surprizing manner. Sometimes the Bodies would move but slowly towards the Tube, sometimes remain a small time suspended between the Glass and the Table on which the Brass Leaf was laid; and sometimes seem to slide along the sides of it without touching.[17]

The general similarity between this passage and Smith's note is obvious. The same experiment is described in Hauksbee's *Physico-Mechanical Experiments* and here the similarity is even closer:

> Having procured a Tube, or hollow Cylinder, of fine Flint Glass, about one Inch diameter, and thirty in length; I rubb'd it pretty vigorously with Paper in my Hand, till it had acquired some Degree of Heat. I then held it towards some Pieces of Leaf-Brass; which were no sooner within the sphere of Activity of the Effluvia emitted by the Tube, but they began to [be] put into brisk Motions, and yielded the following surprizing Appearances.
>
> They would leap towards the Tube, at a very considerable distance from it; nay, I have found, that sometimes the Distance of twelve or more Inches, has not prevented their doing so.
>
> Sometimes they would adhere and fasten to the Tube, settling themselves on its Surface, and there remain quiet: And sometimes they would be thrown off from it with a very great force, even to the distance of six or seven Inches. And not only when they adher'd to the Surface of the Tube, would they thus suddenly and precipitantly be driven from it; but also in their Motion of Ascent towards it, even when they were advanced so far as to touch the Tube, this repellant Force would take place, and hurry them downwards with a great velocity.
>
> And (which still adds to the Wonderfulness of the Phaenomenon) they would often repeat this alternate Rising and Falling; the Attractive and Repulsive Forces (whatever they are) exerting themselves as it were by turns; the one drawing up, and the other beating down these light Bodies; and that for several times one after the other.
>
> Neither is this all the Variety which the Phaenomenon afforded: For sometimes they would move but slowly towards the Tube, sometimes they would remain a small time suspended between the Tube and the Table on which they were first laid; and at other times (which is no less strange than the former)

[17] *Phil. Trans.*, 1706, *25*: 2327–2328; Hutton's Abridgment, 1703–1712, *5*: 324–325.

they would seem to slide along in the Direction of the Sides of the Tube, and that without touching it.[18]

If the underlined portions of this passage are compared with Smith's note, it will be seen that the similarity extends to the actual phrasing. Can there be any doubt that the "lecture note" was based, not on what Smith saw Dr. Spencer do, but on what he read in Hauksbee's *Physico-Mechanical Experiments*, or, less likely, in the account given in the *Philosophical Transactions*? If we bear in mind the statement made by Franklin in the *Autobiography* that "Spence's experiments were imperfectly perform'd, as he was not very expert," we shall probably be justified in concluding that Spencer in his lecture attempted the experiments with the tube and leaf-brass, but with so little success that Smith turned to Hauksbee's work for a description of what *ought* to have happened, and William Black did not think the matter worth mentioning.

One thing seems certain: Smith's notes do not enable us to form any idea of the "electrical" experiments Spencer showed Franklin in 1743, beyond the fact that they were performed with a glass tube and some leaf-brass. We cannot even say with certainty that Spencer himself had any clear notion of their electrical character, for Smith's use of the term "electrical attraction" may have been taken from Hauksbee. What we do learn from both Smith's and Black's notes is that Spencer does not seem to have appreciated the electrical nature of the suspended-boy experiment. Had he done so, he would surely have included a demonstration of the "attraction" for leaf-brass acquired by the boy's face and hands when a rubbed tube was held near his feet, the form which the experiment almost invariably took. Instead, he detaches the suspended-boy experiment from the leaf-brass experiment and treats it as an illustration of the phenomena of "fire." Nor does Smith refer to this experiment as "electrical" — he would of course find nothing about it in Hauksbee's work, since it was not until 1729 that this particular experiment was performed.[19] It is perhaps significant that Franklin himself seems never to have performed the suspended-boy experiment,[20] even in the first flush of enthusiasm following the receipt of the electric tube; it is hardly likely that Collinson described the experiment in his letter, since by that time it was rather out of date, but one would expect Franklin to have recalled it if it had ever been shown to him as an *electric* experiment. On the whole I am inclined to think that *electricity* entered very little, if at all, into Spencer's "philosophical course" or into the demonstration Franklin saw in 1743.

With regard to Franklin's statement in the *Autobiography* that "Dr. Spence" showed him some "electrical" experiments, it must be borne in mind that this was written some time after the event, when Franklin would naturally describe them as *electric*, even though he did not realize they were when he first saw them.

[18] F. Hauksbee, *Physico-Mechanical Experiments on Various Subjects. Containing an Account of Several Surprizing Phaenomena touching Light and Electricity, producible on the Attrition of Bodies.* 2nd Edition (London, 1719), pp. 53–54.

[19] By Stephen Gray. *Phil. Trans.*, 1731, *37*: 39–40; Hutton's Abridgment, 1724–1734, *7*: 459.

[20] Cohen, *loc. cit.*, p. 11.

The inconsistencies in Franklin's statements regarding his introduction to electricity can be cleared up if we accept the following sequence of events. In 1743 Franklin saw Spencer attempt some experiments with a glass tube and some leaf-brass; they naturally interested him, but meant little to him at the time. In 1746 the Library Company of Philadelphia received Collinson's gift of an "electric tube" with directions for its use. Franklin now realized for the first time the true character of the experiments he had seen attempted three years earlier, and so during the next few months "had little leisure for anything else" but electric experiments. When writing the *Autobiography* he can say with perfect truth that the first time he saw such experiments was when he was in Boston (in 1743, not 1746). When writing to Michael Collinson much later he can with equal truth say that the "first Notice" he had of electricity was from Peter Collinson in 1746. Spencer introduced Franklin to an experiment with a glass tube and leaf-brass; Collinson introduced him to *electricity*.

Franklin and Kinnersley†

By J. A. L. Lemay *

THE *Gentleman's Magazine* for January, 1750, published a short article entitled: " By a Number of Experiments, lately made in Philadelphia, several of the principal Properties of the Electrical Fire were demonstrated, and its effects shewn." Although this article has been considered Benjamin Franklin's first published piece on electricity, I am able to show that his first published writing on electricity appeared nearly a year before this. Also I am now able to offer proof of the perceptive suggestion of I. Bernard Cohen that the *Gentleman's Magazine* article is actually a copy of an early outline of the Reverend Ebenezer Kinnersley's lectures on electricity.[1]

Ebenezer Kinnersley (1711–1778) was a well-known and respected minister, educator, and scientist of colonial America. His early fame as an opponent of the enthusiastic tendencies of the Great Religious Awakening was later eclipsed by his renown as a lecturer and theorist in the science of electricity. He was a neighbour, friend, and co-experimentor of Benjamin Franklin. The full scope of Kinnersley's achievements has not been realized. I do not propose to treat his entire career at this time, but merely to point out those aspects of his work in electricity which contribute to our knowledge of Franklin. Of concern must be the charge that Franklin, to some extent, stole Kinnersley's theories. I will, in part, show that there is little reason to believe this, but there is indisputable proof that this was a well-accepted minority belief in colonial America.

Previously it has been thought that Kinnersley's first lecture on electricity was that which was announced in the *Pennsylvania Gazette* for April 11, 1751. Actually he gave his lectures nearly two years before this. A series of advertisements and editorial notes in the May–June, 1749, issues of the *Maryland Gazette* prove that Kinnersley was then lecturing in Annapolis. The probable reason for their not being previously noticed is that none of the notices in the *Maryland Gazette* gives Kinnersley's name. There can be no mistake about their origin since they are practically identical with later advertisements known to be his. An additional proof is that Kinnersley was in Annapolis at this time, for he was a guest at the Tuesday Club in Annapolis, Maryland, on May 16 and June 13, 1749.[2]

† This paper in another form was prepared for a seminar at the University of Maryland.

* The University of Pennsylvania.

1 I. Bernard Cohen, *Benjamin Franklin's Experiments* (Cambridge, Massachusetts, 1941), p. 89.

2 Records of the Tuesday Club are in bound manuscript form in the Maryland Historical Society. The Kinnersley references are on pages 135 and 138-139.

Kinnersley's lectures are actually the joint product of Kinnersley and Franklin. In his *Autobiography* Franklin points out that Kinnersley was one of the principal collaborators in his electrical experiments. Since Kinnersley lacked employment at the time, Franklin encouraged him

> to undertake showing the experiments for money, and drew up for him two lectures, in which the experiments were rang'd in such order, and accompanied with such explanations in such method, as that the foregoing should assist in comprehending the following. He procur'd an elegant apparatus for the purpose, in which all the little machines that I had roughly made for myself were nicely form'd by instrument-makers. . . .[3]

It cannot be positively known to what extent the published outline was written or altered by Kinnersley, but I feel that his influence is in the experiments. A survey of Kinnersley's advertisements over the years shows that he constantly changed their order, re-worded them, and added others. As Kinnersley was an extremely careful conscientious person, it is not likely that he would have begun his tour of the Southern colonies until he had prepared every detail of his lectures and rehearsed them until they met his standards.

It may at first be thought that the appearance of the article in the *Gentleman's Magazine* proves that Franklin was the author of the lectures as they are printed in the *Maryland Gazette,* for they probably came to the editor of the *Gentleman's Magazine* through the hands of Peter Collinson, who had received them directly from Franklin. Nevertheless, the copy of the lectures Franklin sent to Collinson was not his own, but a copy from the *Maryland Gazette.* I. Bernard Cohen has pointed out that *lightning* in the *Gentleman's Magazine* article is spelled without the *e*: Franklin spelled it with an *e*; Kinnersley did not.[4] An additional consideration is that in both the *Maryland Gazette* and the *Gentleman's Magazine,* only six of the experiments of the first lecture are given, while the complete number in the second lecture is listed. It is beyond the bounds of credence to suppose that the outline for the lectures would be cut, independently, in the same place. The *Gentlemen's Magazine* article is a slightly altered copy of the advertisements which appeared in the *Maryland Gazette* for May 10, 17, and 24, 1749.

Kinnersley's lectures excited great wonder from the spectators and proved quite rewarding financially. Jonas Green, the editor of the *Maryland Gazette,* inserted in his edition of May 31, 1749:

> We hear, that the gentleman who exhibits the electrical experiments, designs, before he leaves this place (which will be in a few days), to fire spirits of wine, and discharge his battery of eleven guns, by an electrical spark, that shall first pass thro' the water from Mr. Hill's point to Mr. Carroll's; which is supposed to be about a quarter of a mile.

And in the *Maryland Gazette* of June 14, 1749, Jonas Green tells the outcome of the experiment:

[3] Albert Henry Smyth (ed.), *The Writings of Benjamin Franklin* (10 vols., New York, 1905-1907), I, 417-418.

[4] Cohen, *op. cit.*, p. 89.

On Friday last, the gentleman who has exhibited the electrical experiments in town, removed his machine over to the south side of our creek; and having set some spirits of wine, in a small vessel, on a table on the north side, he caused a spark of electrical fire to dart across in an instant, through 200 yards of water, which set the spirits in a blaze the first attempt, and several times afterwards; and discharged a battery of eleven guns, to the surprise and great satisfaction of the spectators. His experiments are all of them very curious and entertaining, and have given general satisfaction to all who have seen them here. He intends the first opportunity for Norfolk, and other parts of Virginia.

We may presume that Kinnersley left for Virginia in the latter part of June, 1749. Thus far I have been able to find no records from Virginia which positively prove that he lectured there. But I believe that the following news item (evidently reprinted from the *Virginia Gazette*) [5] which appeared in the *Maryland Gazette* for October 18, 1749, is about Kinnersley:

Williamsburg, August 31.

We have receiv'd the following Certificate from Suffolk: Suffolk, August 18, 1749.

The gentleman who has been entertaining us with a Course of very curious Electrical Experiments has also applied the Electrical Fire to the human frame, with remarkable and speedy success, in curing the tooth ache, pains in the head, deafness, pains in the limbs, which had been so violent as to take away the use of them, pain in the stomach, swelling of the spleen, sprains, relaxation of the nerves, &c. The most remarkable are the two following instances, *viz.*

One Samuel Miller, who for three years past could not lift his hand above his head without putting his shoulder out of joint, by a few applications of the electrical fire has met with a perfect cure.

An negro boy, about sixteen years of age, who had always been so deaf as scarcely to hear the loudest sounds, has by the same means been brought to hear, when spoke to in a common tone of voice.

We the subscribers thought proper to give this information to the public, that others, who may have the opportunity, might be encouraged to make further trial of this wonderful remedy.

Robert Brown,	William Webb,	Robert Cook,
David Meade,	John Watson,	John Marlow,
Lemuel Riddeck,	John Wright,	Alex. Cairnes.

The only Southern newspaper published at this time which has known extant copies is the *South Carolina Gazette.* Kinnersley did not advertise in this *Gazette.* The *South Carolina Gazette* for May 21 to May 28, 1750, reproduces (probably from the *Gentleman's Magazine*) the syllabus of experiments for Kinnersley's lectures; therefore, it seems that the people of Charleston could not have been acquainted with Kinnersley's lectures

[5] There are no extant copies of the *Virginia Gazette* for 1749.

This same article was also printed in the *Pennsylvania Journal* for December 5, 1749, and in the *Boston Evening Post* for January 1, 1750.

before then. After lecturing in Virginia, Kinnersley probably returned to Philadelphia for the winter.

By February, 1750, Kinnersley was in Philadelphia, for on February 4, 1750, Franklin wrote to Collinson,[6] telling him of the experiments that he and Kinnersley had just performed. After lecturing in Philadelphia, Kinnersley left for his well-known tour of the Northern colonies. He lectured in Boston, Newport, New York, Philadelphia, and St. John's, Antigua, until he was recalled by Franklin to take the post of Master of the English School at the College and Academy of Philadelphia on July 10, 1753.[7] From December 27, 1753, when his advertisement of the " electrical experiments " appeared in the *Pennsylvania Gazette*, until March 2, 1774, when his last advertisement appeared, Ebenezer Kinnersley lectured at least several times a year.

By 1753, when Kinnersley returned to Philadelphia to make his lectures there something of an institution, he was a well known theorist and experimentor in electricity throughout the colonies and in the West Indies. Franklin's fame as an electrician had to come to America from Europe, where the first edition of his book *Experiments and Observations* . . . was published in 1751. It is partially this consideration that leads me to accept the opinion that when Franklin's name became inextricably associated with electricity — and for the very same experiments and theories which so many of the colonists had seen performed by Kinnersley — there existed a widespread rumor that Franklin had plagiarized Kinnersley.

It was in the issue of the *American Magazine*, of October, 1758, in an article by William Smith, " Account of the College and Academy of Philadelphia " (pp. 630–640), that the charge that Franklin had stolen Kinnersley's theories first appeared in print:

> [Ebenezer Kinnersley] is well qualified for his profession; and has morover great merit with the learned world in being the chief inventor . . . of the electrical apparatus, as well as author of a considerable part of those discoveries in electricity, published by Mr. Franklin to whom he communicated them. Indeed Mr. Franklin himself mentions his name with honor, tho' he has not been careful enough to distinguish between their particular discoveries. This, perhaps he may have thought needless, as they were known to act in concert, But tho' that circumstance was known here, it was not so in the remote parts of the world to which the fame of these discoveries have extended.

The Reverend William Smith, Provost of the Philadelphia Academy, was in a position to know. But he was by this time a political enemy of Franklin. Therefore, it is not his statement which is of interest, but the fact that he

[6] This letter was printed by A. H. Church, *The Royal Society: Some Account of the Letters and Papers of the Period 1741-1806 in the Archives* (Oxford, 1908), pp. 10-12, and reprinted in major part in Cohen, *op. cit.*, pp. 93-94.

[7] Minutes of the Trustees of the College, Academy, and Charitable Schools, I, 32-33. This is quoted by Thomas Harrison Montgomery, *A History of the University of Pennsylvania* (Philadelphia, 1900), pp. 170-171.

felt that he could make this charge. It supports my view that Kinnersley's name was more associated with electricity by the general public than was Franklin's until the latter's European fame spread to America. Naturally some portion of the population suspected that the electrical experiments and theories were really Kinnersley's. This suspicion was, Kinnersley himself says, false.

An evaluation of the contributions of the various members of the performers of the " Philadelphia Experiments " was given by Kinnersley in a letter refuting William Smith's charges. The letter was published in the *Pennsylvania Gazette* for November 30, 1758, and, to the best of my knowledge, has never before been reproduced:

> To the Author of the Account of the College and Academy of Philadelphia, published in the *American Magazine* for October, 1758.
>
> Sir,
>
> I was very much surprised and concerned to see the Account you have been pleased to give of my electrical Discoveries, in page 639 of the *American Magazine*. If you did it with a view to procure me esteem in the learned world, I should have been abundantly more obliged to you, had it been done, so as to have no tendency to depreciate the merit of the ingenious and worthy Mr. Franklin, in the many curious and justly celebrated discoveries he has made in electricity. Had you said that, being honoured with Mr. Franklin's intimacy, I was often with him when he was making experiments, and that new discoveries were sometimes made when we were together, and at other times some were made by myself at home, and communicated to Mr. Franklin, this would have been really true, though it is what I never desired to have published. But to say, ' That I am the author of a considerable part of those discoveries in electricity, published by Mr. Franklin '; the expression, from whomsoever you might have the intelligence, appears too strong; it may be understood to comprehend more than is strictly true, and therefore I thought myself obliged to take this public notice of it. If you will please, Sir, to examine what Mr. Franklin has published on electricity, I think you will no where find that he appropriates to himself the honour of any one discovery; but is so complaisant to his electrical friends, as always to say, in the plural number, *we* have found out, or, *we* discovered, &c. As to his not being careful to distinguish between the particular discoveries of each; this perhaps was not always practicable; it being sometimes impossible to recollect in whose breast the thought first took rise, that led to a series of experiments, which at length issued in some unexpected important discovery. But had it been always practicable to distinguish between the particular discoveries of each, it was altogether unnecessary; as, I believe, none of Mr. Franklin's electrical friends had the least thought of ever appearing as competitors for any of the honours that they have beheld, with pleasure, bestowed on him, and to which he has an undoubted right, preferable to the united merit of all the electricians in America, and, perhaps, in all the World.
>
> I am, Sir, Your most obedient humble Servant,
>
> Ebenezer Kinnersley

It is likely that this letter of Kinnersley's led Franklin to try to credit each of his fellow experimentors with their ideas. Franklin noted in his copy of the *Experiments and Observations* which of the discoveries were performed by someone other than himself, and these notes appeared in the fourth English edition of the book in 1769 and thus in subsequent editions.

Although the book enjoyed a wide European circulation, it was not printed in America until 1941, and Franklin's notes of credit evidently did little to stop the rumor. Jonathan Boucher, in his book *A View of the Causes and Consequences of the American Revolution* (London, 1797), wrote an attack on Franklin:

> Franklin's enemies, however bitter, have seldom been found so wanting in truth and justice as to deny him great merit in his philosophical character: it was in Philadelphia chiefly, if not solely, and by his friends, that he was charged with having stolen from an Irish gentleman, of the name of Kinnersley, many of his useful discoveries respecting electricity. How truly he was, or was not, the discoverer of the electrical nature of lightning, I cannot, amid such a variety of contradictory evidence, take upon me to determine; but common justice requires that I should acknowledge, that, in his day, no man contributed more to excite and foster a spirit for investigation and experiment; and that he first effectually practised, what Lord Verulam first conceived and recommended, *viz.* the stripping philosophy of her uncouth scholastic garb and rendering her the companion and friend of all orders of men.
>
> Tender and cautious as I am, and ought to be, of bringing a charge of plagiarism against a man who can no longer vindicate himself, I cannot help observing, that though I certainly have often heard the allegation used against Dr. Franklin in America, and though it was set down as it now stands in this place soon after my hearing it in Philadelphia, from a gentleman who was well acquainted with both the parties; it must strike everyone as amounting almost to a direct refutation of the charge, that Kinnersley does not appear to have claimed any share in a discovery to which Dr. Franklin publicly avowed his own claim, but this successful plagiarism, admitting it to have been one, is not the only instance of it's kind imputed to the Doctor. . . .[8]

In his autobiography, *Memoirs of a Life Chiefly Passed in Pennsylvania* (Harrisburgh, 1811, p. 16) Alexander Graydon recalled his former school master, the Reverend Ebenezer Kinnersley, as a " large, venerable looking man, who, whether truly or not, has been said to have had a share in certain discoveries in that science of which Doctor Franklin received the whole credit."

In *Memoirs of the Late Dr. Benjamin Franklin* (London, 1790, pp. 10-11), Mr. Wilmer reveals that he knows of the rumor but makes an accurate appraisal when he writes:

> In his experiments, it is said, he was assisted by the Rev. Ebenezer Kinners-

[8] Pp. 438-439. I suppose the " gentleman " who was well acquainted with both parties was William Smith. However, Boucher's statement that he ". . . often heard the allegation used against Dr. Franklin in America . . ." is very interesting.

> ley, Professor. . . . He (Franklin) undoubtedly improved by the experience of this ingenious gentleman; but his own sagacious and active mind led him on to discoveries that will immortalize his name. . . .

William Barton, in *Memoirs of David Rittenhouse* (Philadelphia 1813, p. 155), wrote:

> This venerable and worthy man, who was a clergyman of the Baptist Church, was a very eminent Electrician. In this branch of philosophy, he was an able lecturer and ingenious experimentalist; and perhaps to no person – at least in America – were his contemporaries more indebted, than to him, for the light which he shed, at a very early day, on this interesting and pleasing science.

This last statement is true only if one assumes that Reverend Barton is speaking of spreading knowledge of electricity among the people, for Kinnersley well knew that Franklin was his teacher in electricity. His letters to Franklin invariably express the relationship of a gifted student to his teacher: typical is his statement (letter to Franklin dated March 12, 1761), ". . . I should be glad to see these phenomena better accounted for by your superior and more penetrating genius. . . ."

There remains some doubt that Franklin awarded to Kinnersley all the credit due him. Kinnersley was extremely honest and had the greatest integrity. What he had in mind when he wrote that it was ". . . sometimes impossible to recollect in whose breast the thought first took rise, that led to a series of experiments . . . ," we cannot, on the extant evidence, know. Certainly none of the credits that Franklin gives in his book on electricity suggests that anyone other than himself was responsible for a "series of experiments." I would speculate that Kinnersley felt that he deserved such credit; why else include that sentiment? Nevertheless, Franklin was the master, and Kinnersley thought Franklin had an "undoubted right" to the honors he received.

An Appraisal of the Origins of Franklin's Electrical Theory

*By Bernard S. Finn**

IT HAS BEEN ASSERTED that Benjamin Franklin owed his achievements in electricity in large part to his isolation, to the fact that he was not contaminated by European prejudices. As early as 1759 Pieter van Musschenbroek suggested that "you would go on making experiments entirely on your own initiative and thereby pursue a path entirely different from that of the Europeans, for then you shall certainly find many other things which have been hidden to natural philosophers throughout the space of centuries."[1] But even if one grants the plausibility of such a statement, the questions remain: in what way was he isolated, what prejudices was he freed from, and how did this actually affect his scientific development?

In this paper I shall attempt to show that Franklin's acquaintance with European developments was severely limited and that this was probably an important factor in the development of his views. But it is interesting to note that not only was he unprejudiced by certain European theoretical approaches, he was also ignorant of some important experimental evidence; and the two combined their effects in a most fortuitous manner.

Benjamin Franklin's entrance on the electrical scene in the spring of 1747 occurred at a critical moment. The Leyden jar, which dramatically increased the amount of electrical energy available at one time by two or three orders of magnitude, had just been discovered, and the friction electric machine was rapidly being developed. Almost from the beginning, therefore, Franklin knew of electricity in the form of violent shocks and easily obtained sparks.[2] The importance of this can be understood better by considering the theoretical views of his European predecessors, who had been trained with a glass rod and a piece of silk.

It is not surprising that these early electricians were primarily concerned with the phenomena of attraction and repulsion, since the most remarkable and unique characteristic of an electrified body was that it attracted and (to the close observer) then repelled small pieces of paper, threads, and other light bodies. To explain electricity was to explain these phenomena. J. T. Desaguliers, Benjamin Martin, Pieter van

* Smithsonian Institution.

[1] Quoted in L. W. Larabee, ed., *The Papers of Benjamin Franklin* (New Haven: Yale Univ. Press, 1959—), Vol. VIII, p. 329. This translation is taken from I. B. Cohen, ed., *Benjamin Franklin's Experiments* (Cambridge, Mass: Harvard Univ. Press, 1941), p. 71.

[2] Franklin's first experiments were performed with a tube, but by the time of his first detailed letter to Peter Collinson, 25 May 1747, he was using a machine and a Leyden jar. *Franklin Papers*, Vol. III, pp. 132, 134.

Musschenbroek, Abbé J. A. Nollet—among the most prominent electricians of the early 1740's—did this by postulating an electrical effluvium that could swirl out from the rubbed body to impinge upon and in some manner affect the motion of light objects.

A subsidiary question, related to this first problem, concerned the relationship of electricity to other forms of matter, most notably to fire and light. The small sparks obtainable even from a rod seemed to indicate some connection. Even more spectacular were the experiments of Francis Hauksbee, performed early in the century but repeated or reported by many later electricians: an evacuated tube or globe would glow vividly when rubbed in the dark. The Abbé Nollet was only typical in supposing that arguments of simplicity demanded that fire and electricity were merely different manifestations of the same basic element, and the term "electrical fire" was widely used.

The second major problem was how to explain the Leyden jar. In its initial form the jar consisted of a bottle containing water (a condenser effect occurred between the conductive water inside and the hand holding the jar on the outside, with the glass acting as a dielectric), and it seemed that the glass bottle helped to constrain the electric virtue in the water. Further experimentation soon showed that all that was necessary were thin metal coatings on both sides of the glass; the bottle shape was immaterial. The theoretical difficulties of this were so great that Nollet was the only European to attempt an explanation. He supposed that the effluvia accumulated on one side and, upon discharge, somehow were released to penetrate through the glass, an obviously unsatisfactory solution.

The limited nature of Franklin's European contacts is indicated immediately if we consider how he treated action at a distance. The fact of the matter is that he ignored it. Nowhere in his writings is there any explanation of attraction and repulsion or the need for having one or even the desirability of not having one. He described the electrical atmosphere which he supposed surrounded charged bodies and which was quite similar to the effluvia of his contemporaries, but for Franklin this atmosphere was merely surplus fluid with boundaries quite close to the object. It did not extend far enough to explain most observed cases of attraction or repulsion, and it was not used by him to do so.[3]

For Franklin, the active principle was contained in his particles of electric fluid: they were attracted to matter and repelled each other. They permeated all matter, ordinarily in an equilibrium condition. When they moved the result was conduction. They could also be used to explain induction and the Leyden jar. All this was so good, in fact, that when Franklin encountered repulsion between two objects which had less than their normal store of electricity he was, in effect, flabbergasted. It is important to note that he knew of the experimental evidence for this repulsion at least as early as April 1749 and probably in 1748. His only comment was that it "surprizes us, and is hitherto not satisfactorily accounted for."[4] That he did not forget the problem is evi-

[3] Cohen indicates that Franklin used the electric atmosphere to explain repulsion between positively charge bodies, in *Franklin and Newton* (Philadelphia: American Philosophical Society, 1956), p. 468. My own interpretation of the several times Franklin discusses the atmosphere is that this is not the case; furthermore, Franklin's atmosphere was confined close to the charged body, much closer than the repulsive force.

[4] Letter to Collinson, 29 April 1749, in *Franklin Papers*, Vol. III, p. 363. Franklin states that this letter describes experiments performed in 1748.

denced by the following remarks, made in 1755: "Bodies electrified negatively . . . repel each other (or at least appear to do so, by a mutual Receding) as well as those electrified positively, or which have electric Atmospheres."[5] There was no theoretical speculation, and there were no detailed experiments that might have been designed to probe the phenomenon closer. Furthermore, when Franz Æpinus (following the lead of Giovanni Beccaria) altered Franklin's theory in 1759 by allowing the particles of ordinary matter to carry repulsive charges,[6] Franklin praised the book in general but made no mention—at least none that survives—of the crucial modification.[7] Notably, in 1762 he did not have recourse to Æpinus in explaining the spinning motion resulting when a stick was balanced in the middle and two horizontal pins facing opposite directions (perpendicular to the stick) were electrified. If they were electrified positively, he hypothesized, the air opposite the points was also positive (because of ease of discharge from the point) and repulsion took place. If the pins were electrified negatively, the air behind the points would also be negative, but motion would take place by means of attraction between the negative heads of the points and the neutral air in front.[8] In other words, the electrical force rested exclusively in the electrical fluid, not in common matter.

There is also no evidence in his early writings that Franklin was at all familiar with the evacuated experiments of Hauksbee. As mentioned above, he could have read about them in several sources, and their spectacular nature undoubtedly would have made an immediate appeal to him. Partly as a result of this ignorance, Franklin consistently refused to commit himself to any identification of electricity with fire or light.

Franklin's first surviving mention of the Leyden jar occurred in his letter to Peter Collinson dated 25 May 1747. A detailed explanation of its action was presented in July, and he was in good company in supposing that the electricity "crowded" itself into the nonelectric contents of the bottle (water, lead shot, etc.).[9] Since the jar could be discharged merely by connecting the wire sticking out of the top of the bottle to the coating on the bottom, it seemed clear to him that while part of the contents (at the top) might have a surplus of electricity, the other part (at the bottom) had a deficit. It was as if a high-pressure region and a vacuum existed side by side in the same chamber. He marvelled at this remarkable phenomenon:

> So wonderfully are these two States of Electricity, the *plus* and *minus* combined and ballanced in this miraculous Bottle! situated and related to each other in a Manner that I can by no Means comprehend! If it were possible that a Bottle should in one Part contain a Quantity of Air strongly comprest, and in another Part a perfect Vacuum; We know the Equilibrium would be instantly restored *within*. But here we have a Bottle, containing at the same Time a *Plenum* of Electrical Fire and a *Vacuum* of the same Fire; and yet the Equilibrium cannot be restored between them but by a Communication without! Tho' the Plenum presses violently to expand, and the hungry vacuum seems to attract as violently in Order to be filled.[10]

[5] Franklin, "Electrical Experiments," 14 March 1755, *ibid.*, Vol. V, p. 516.

[6] F. U. T. Æpinus, *Tentamen theorae electricitatis et magnetismi* (St. Petersburg, 1759).

[7] Letter to Cadwallader Colden, 26 Feb. 1763, *Franklin Papers*, Vol. X, p. 204; letter to Ezra Styles, 29 May 1763, *ibid.*, p. 266; letter to John Winthrop, 10 July 1764, *ibid.*, Vol. XI, p. 254; letter to Barbeu Dubourgh, 10 March 1773, in A. H. Smyth, ed., *The Writing of Benjamin Franklin*, 10 vols. (New York: Macmillan, 1906), Vol. VI, p. 25.

[8] Letter to Ebenezer Kinnersley, 20 Feb. 1762, *Franklin Papers*, Vol. X, pp. 45–46.

[9] Letter to Collinson, 28 July 1747, *ibid.*, Vol. III, p. 157.

[10] *Ibid.*, pp. 158–159.

By 1749 he realized that the bottle's contents were unnecessary, that a pane of glass coated on both sides would do equally as well, and that the positive charge could be on either side of the glass. He offered no detailed explanation for how the electrical particles arranged themselves, but he did enunciate a principle of conservation, stating that the amount charging one side of the glass was exactly balanced by the amount leaving the other side: "Glass . . . has, within its Substance always the same Quantity of Electrical Fire; and that, a very great Quantity in Proportion to the Mass of the Glass."[11] He tested his views with some excellent experiments. In one he poured the water from the jar, finding that the charge was left behind (William Watson later repeated this experiment unsuccessfully, in part because he failed to understand Franklin's line of reasoning[12]).

In 1750 Franklin imagined that the electrical fluid, or electrical fire, could be accumulated over the conductive coating on one side *A* of the glass, where it would repel the particles of electric fluid from the opposite glass surface *B* to the coating on side *B*, "whence they are discharged." The fluid on coating *A* could then enter into surface *A* of the glass: "But when this is done, there is no more in the glass, nor less than before; just as much having left it on one side, as it received on the other."[13] By locating the charge at the surface of the glass, Franklin had a theory that sounds quite modern. Unfortunately, we might say, he went on to describe what he meant by "surface":

> By the Word *Surface* in this Case, I do not mean mere Length and Breadth without Thickness: But when I speak of the upper or under Surface of a Piece of Glass, the outer or inner Surface of the Phial, I mean Length, Breadth and half the Thickness; and beg the Favour of being so understood.[14]

This obviously left him with the problem of explaining why the electricity did not shift from one side to the other through the middle. He then suggested that most of the glass acted like a sponge in soaking up electrical fluid:

> But I suppose farther, that in the Cooling of the Glass, it's Texture becomes closest in the Middle, and forms a Kind of Partition, in which the Pores are so narrow, that the Particles of the Electrical Fluid, which enter both Surfaces at the same Time, cannot go thro' or pass and repass from one Surface to the other, and so mix together.[15]

Franklin apparently tested his theory soon after writing the above, although there is no evidence that he mentioned it until 1755, when he reported: "I ground away five-sixths of the thickness of the glass, from the side of one of my phials, expecting that the supposed denser part being so removed, the electric fluid might come through the remainder of the glass, which I imagined more open; but I found myself mistaken."[16] He was left with no theoretical alternative: "I am now, as much as ever, at a loss to know

[11] Letter to Collinson, 29 April 1749, *ibid.*, p. 355.

[12] William Watson, "Notice of Franklin's Experiments," 11 Jan. 1750, *ibid.*, p. 458. Franklin poured the water into an insulated container, then tested it—with negative results. Watson tested the water as it was poured and obtained positive results in the form of a spark; the stream, of course, formed a conductive path back to the jar, as Franklin later pointed out.

[13] Letter to Collinson, 29 July 1750, *ibid.*, Vol IV, p. 27.

[14] *Ibid.*

[15] *Ibid.*, p. 28.

[16] Letter to John Lining, 18 March 1755, *ibid.*, Vol. V, p. 523.

how or where the quantity of electric fluid, on the positive side of the glass, is disposed of."[17] There is no evidence known to me that Franklin ever worried further about this problem or that he performed any additional experiments to investigate it. Nor is there any evidence that knowledge of it prompted him to modify his theory at any stage.

If Franklin had taken seriously his inability to correlate the properties of his electric fluid with an explanation of any of the problems raised above, his entire theoretical structure would have been severely shaken, if not demolished. To explain action at a distance he might have taken the path indicated by his contemporaries and somehow allowed his electrical fluid to act as an ether or effluvium and to produce the effects of action at a distance, but it is difficult to see how he could have avoided painting a confusing and inconsistent picture. To have included repulsion between negatively charged objects in his theory, he could have recognized two electrical fluids or postulated repulsion between particles of ordinary matter; either would have severely altered the simple concept of a single electrical matter as the sole seat of electrical forces. The Leyden jar was at least as great a problem: either the glass soaked up electrical fluid (like a conductor) or it did not (like an insulator). There was something obviously wrong with having it do both at the same time. These problems, neglected by Franklin, had been faced by his predecessors and contemporaries—much to their sorrow. For the state of the art in the mid-eighteenth century was such that it was virtually impossible to answer them in a satisfactory manner and at the same time to explain the various phenomena of conduction and induction.

Hindsight tells us that Franklin was fortunate in avoiding these questions, even if he had no good reason at the time. A need to explain action at a distance was already becoming intellectually less and less important—a philosophical trend to which Franklin contributed. Repulsion between negative charges would be "explained" in Franklin's lifetime by Beccaria and Æpinus (and also by two-fluid theories) in a manner that was obviously quite unpalatable to Franklin but quite acceptable to others. The third item—the Leyden jar—would only be satisfactorily described after the experiments of Faraday on dielectric polarization in the mid-nineteenth century, and Franklin's theory could easily accommodate this new information.

Franklin's fortuitous neglect was a result, I think, of an early and strong commitment to his fluid theory; and this, in turn, was directly connected to his isolation from the European scientific community. A better feeling for this can be obtained from a detailed consideration of the development of his views.

It is difficult to assign precise origins to Franklin's electrical speculation, primarily because he made virtually no references to predecessors in his writings. As is well known, in his autobiography he mentioned that his first exposure to electrical experiments was through a Mr. Spence (Adam Spencer). Present opinion is that this was undoubtedly a very casual thing[18] and that Franklin's interest was really stimulated only after Peter Collinson had sent the Library Company a glass tube, probably some time in 1746 (on 28 March 1747 Franklin wrote that he had been engrossed in electrical experiments "for some months past"[19]). Cohen suggests that Franklin may have

[17] *Ibid.*

[18] N. H. deV. Heathcote, "Franklin's Introduction to Electricity," *Isis*, 1955, *46*:29–35. Cohen, *Franklin and Newton*, p. 435.

[19] Letter to Collinson, 28 March 1747, *Franklin Papers*, Vol. III, p. 119.

had some acquaintance with the electrical writing of Desaguliers and Hauksbee through one of two or three semi-popular sources available to him; at the same time he argues that the major influence on Franklin in this early period was the work of William Watson.[20]

I think it is possible to go a step beyond this and infer that in the critical period up through 1748 and quite likely through 1750, Franklin did not read or at least was not influenced by the early electricians (before Watson). They had laid great stress on explaining attraction and repulsion; they also had been at pains to describe repulsion between like charges of both kinds of electricity. It is difficult to see how Franklin could have avoided at least mentioning these subjects if he had known of the prior attention given them. Additionally, there was the glow of the electrified evacuated glass container. Finally, the early terminology was not used by Franklin. In the 1730's Desaguliers (typically) employed the terms "electric effluvia" and "virtue," and he used C. F. Dufay's words in differentiating between the two types of electricity as "vitreous" and "resinous." These terms simply do not appear in Franklin, even by way of clarification.

On the other hand, a close look at his writings gives good cause to believe that Franklin depended very heavily indeed upon the works of Watson. Franklin reported his electrical researches in a series of letters, most of them addressed to Peter Collinson in London. In the first letter, dated 28 March 1747, Franklin mentioned merely that he was engaged in electrical experiments. Details followed on 25 May. Sometime in March or April he received a package (sent from England in March) containing copies of Watson's *Experiments and Observations Tending to Illustrate the Nature and Properties of Electricity* and its *Sequel* (London, 1746), and Benjamin Martin's *An Essay on Electricity* (Bath, 1740).[21] It is interesting to compare Franklin's May terminology with what was available in these works. He referred to the electrical material exclusively as "electrical fire" and "fire"; these terms appeared in both Watson and Martin (though certainly not exclusively there). "Points," "communicate," "circulate," and "electric atmosphere" appeared in Watson. Even "afflux" and "efflux," which Franklin mentioned once, were used by Watson in discussing Nollet (Franklin used them in a different sense). In describing the Leyden jar, Franklin used the term "vial" eight times, "phial" once, and "bottle" once. "Vial" comes from Watson (I have seen it no place else), and "phial" from Watson and Martin; "bottle" seems to be a Franklin original.

If Watson's little book was in fact essentially Franklin's sole source in 1747, this would help to explain the ease with which he developed his theory and the strength with which he adhered to it. Watson described the attractive and repulsive powers of electricity, but he did not try to explain them; the electric fire, which could be differentiated from other fire, was apparently held in the pores of the glass; it flowed in a circuit; the electrical machine was a fire pump in direct analogy to an air pump; and he talked of equilibrium. All this would not inevitably lead to the theoretical views of Franklin's May letter. But given such a background, uncluttered by the speculations of Desaguliers and Nollet and the experiments of Hauksbee, there would have been no impediment to supposing that electric fire was a common element held in all matter,

[20] Cohen, *Franklin and Newton*, p. 413.

[21] Mentioned in letter to Collinson, 25 May 1747, *Franklin Papers*, Vol. III, p. 134.

that it could be circulated, and that a body could receive too much (plus) or too little (minus) of it.

These views could only have been reinforced by the receipt—in June or July[22]— of a small book by Benjamin Wilson in which it was stated that "the electric matter was in all bodies, in reciprocal proportion to their densities. . . . Further, when any body was electrified, I called that quantity of electric matter, by which it was electrified, the excess of electric matter in that particular body, above its natural original quantity."[23]

The evidence available indicates that Franklin continued to be limited in his exposure to the works of others during the next two or three years. He probably received Volume 44 (1746) of the *Philosophical Transactions of the Royal Society* in June 1747.[24] This contained interesting accounts of experiments by G. Le Monnier as well as a direct description of the Musschenbroek experiment with the Leyden jar, but there was nothing very interesting theoretically, and the experiments did not raise any new problems.

The next communication of substance by Franklin after July 1747 did not come until 29 April 1749. There is no direct indication of what he may have read in the intervening period, except that he "lately received" Watson's last book.[25] It is reasonable to suppose that he continued to receive copies of the *Philosophical Transactions*, which carried several articles on electricity. The most pertinent of these were by Watson, especially one dated January 1748.[26] Franklin adopted some terms which apparently came from this article, namely "coating," "Hook," "charge," and "discharge" when referring to the Leyden jar, and "prime conductor" when referring to electrical machines.

After 1750 references to various other electrical works crept into his correspondence. But by now Franklin's major commitment and contributions to electricity were finished.

It is important to note here the brief span of Franklin's principal electrical activities. They began late in 1746 or early in 1747; basic concepts (plus, minus, the Leyden jar) had been worked out by summer. Further experiments were performed during the winters of 1747–1748 and 1748–1749 (electrostatic experiments are not easily performed in the humid summers of Philadelphia). In the spring of 1749 he made important modifications to his theory of the Leyden jar and also discussed the electrical nature of lightning. In the summer of 1750 he summarized his theoretical views; he also suggested the lightning experiment. Although he performed occasional experiments and wrote a few letters after this, the main thrust of his electrical activities was ended.

Let us now attempt to draw all of this together. In the first place, Franklin had a good basic hypothesis, developed as a direct extension from Watson: electricity was a conservative electrical fluid in which the particles repelled each other but were strongly attracted to matter. It nicely explained phenomena of conduction and induction and,

[22] Mentioned in letter to Collinson, 28 July 1747, *ibid.*, p. 156.

[23] Benjamin Wilson, *An Essay towards an Explication of the Phenomena of Electricity* (London: C. Davis, 1746), p. x.

[24] Letter from Collinson, 1 June 1747, *Franklin Papers*, Vol. III, p. 141.

[25] William Watson, *An Account of the Experiments Made by some Gentlemen of the Royal Society* (London: 1748). Franklin mentions receiving it in his letter to Collinson, 29 April 1746, *Franklin Papers*, Vol. III, p. 358.

[26] William Watson, "A Collection of the Electrical Experiments Communicated to the Royal Society by Wm. Watson," *Philosophical Transactions of the Royal Society*, 1748, *45*:49–120.

in a sense, the Leyden jar. For the finer points of the Leyden jar, and for the relationship between negatively charged particles, however, it broke down—or would have if Franklin had pushed it.

He did not in large part because he was not pressed to do so. Initially his geographic isolation insulated him from troubling and critical ideas which might have interrupted the formulation of his system. Although his letters were read to the Royal Society, only sketchy summaries appeared in the periodical literature. By the time his book was published in England in 1751 and was translated into French in 1752, two additional important factors were coming into play. One was his own relative withdrawal from the scientific scene; he wrote occasional letters when provoked by inquiries but did not actively engage in any self-criticism. The other factor was the lightning experiment, first performed in France in June 1752; with it his reputation was immediately secure.[27] As a result, Nollet was virtually the only one to offer a dissenting view,[28] and repulsion between negative charges was incorporated into the system as only a minor modification.[29]

Finally, the successes of the fluid theory tended to direct attention into new paths. The problems of the 1740's, if not completely solved, were at least no longer so interesting. The literature of the second half of the century is dominated by applications—bigger and better machines being used to produce bigger and more spectacular sparks that could be used for physical, chemical, and biological experiments.

Franklin was ahead of his time, though probably not by much. The writings of Watson and Martin were too close to the mark, and the inadequacies of other theories were too great. If not by Franklin, a suitable fluid theory would have been devised by someone else—perhaps Watson, or more likely Beccaria or Æpinus. Franklin arrived on the scene at a most propitious moment, and he gained from his isolation in two important ways: he was limited in his knowledge and he was spared criticism. Clearly it is curious to call this an advantage. But Franklin, like many before and after him, was lucky—what he did not know might have confused him, and criticism might have forced him to reconsider the holes in his theory that he had cheerily circumvented.

[27] It is interesting that Joseph Priestley called the identification of lightning and electricity "the greatest discovery which Dr. Franklin made concerning electricity," in his *History and Present State of Electricity* (London: Printed for J. Dodsley, . . ., 1767), p. 170.

[28] Nollet's principal objections, significantly, concerned the nature of electricity and the Leyden jar. On the first, he argued that electrical fire should be considered identical with common fire; on the second, he argued in favor of the electrical fire penetrating through the glass sides of the jar. *Lettres sur l'électricité* (Paris: H.-L. Guerin & L.-F. Delatour, 1753), pp. 49 ff.

[29] As in Prie tley, *op. cit.*, p. 462.

Thomas Robie (1689-1729), Colonial Scientist and Physician

THOMAS ROBIE, " ... the most famous New Englander in science in his day..." (1), was born in Boston on March 20, 1688/89. Although his parents were members of a worthy and respectable class, they were not outstanding individuals. The exact character of ROBIE's early schooling is unknown, but it was sufficient to enable him to enter Harvard with the class of 1708. His behavior in college was that of an average boy (2), and he must have been considered a deserving student, since the grant of forty shillings which the Harvard Corporation voted him on August 22, 1705 (3) was only the first of three awards which ROBIE received while he was an undergraduate.

After he had graduated in 1708, ROBIE taught school in Watertown, Massachusetts for a half year, and then returned to Boston. Like many Harvard graduates of the seventeenth and early eighteenth centuries, he now began to utilize the scientific training which he had received in college by preparing an almanac for the year 1709 (4). ROBIE continued to publish these annual almanacs until 1720. On July 4, 1711, he took his second Harvard degree, that of Master of Arts, and once again the college gave him financial aid when, a month and a half later, the Corporation allowed ROBIE five pounds from the Keayne legacy. He apparently maintained a close if unofficial relationship to the college, and on September 9, 1712, he was voted both the Harvard " Library-

(1) C. K. SHIPTON, *Sibley's Harvard Graduates. Biographical Sketches of Those Who Attended Harvard College in the Classes* 1701-1712 (Boston : Massachusetts Historical Society, 1937), V. p. 452.

(2) *Ibid.*, p. 450.

(3) *Harvard College Records* [Boston : Colonial Society of Massachusetts, 1925-35], (Publications of the Colonial Society of Massachusetts, Volumes XV-XVI, XXI, Collections), I, p. 374.

(4) THOMAS ROBIE, *An Ephemeris... for the Year...* 1709 (Boston : [1708]).

keeper" and a Scholar of the House (5), positions which he held for the academic year 1712/13. As a further mark of approbation, the Harvard Corporation, on April 7, 1713, elected him a "Fellow of the House," as the Harvard Tutors were officially designated. He was also made a Fellow of the Harvard Corporation in July, 1722, but he resigned from both this position and his tutorship the following year when he moved to Salem.

ROBIE is known as one of Harvard's "scientific tutors," but his interests included many subjects in addition to science and the practise of medicine. The notes in his MS Commonplace Book (6) range from excerpts taken from FRANCO BURGERSDIJCK's *Institutionum Logicarum* to the more personal subject of the ethics of sexual relationships. A sampling of the non-scientific contents of this book reveals the following : "Several Opinions of Predestination," an "Alphabet of Prophetick Iconisms," notes on HUGO GROTIUS' *De Jure Belli ac Pacis* and, in particular, many observations on theological and biblical subjects. He had, in fact, seriously considered entering the ministry before his appointment to a tutorship in 1713, and to this end had preached on trial in several towns (7). While he was in Cambridge he continued to preach, at least one of his sermons having been published (8).

ROBIE's scientific interests led him for the most part into the fields of meteorology, astronomy, and medicine. He corresponded with MATTHEW WRIGHT of Crew and with the Reverend WILLIAM DERHAM about various astronomical and meteorological subjects; it was at the latter's request that he kept a meteorological record of Cambridge weather conditions for the years 1715 through 1722 (9). After ROBIE's death in 1729, ISAAC GREENWOOD sent this material to DERHAM, who published parts of it with his own conclusions (10).

---

(5) *Harvard College Records*, II, p. 405.

(6) At the Massachusetts Historical Society.

(7) SHIPTON, *op. cit.*, p. 451.

(8) THOMAS ROBIE, *A Sermon Preached in the College at Cambridge* (Boston : Printed by S. KNEELAND, 1721).

(9) WILLIAM DERHAM, "An Abstract of the Meteorological Diaries Communicated to the Royal Society," Royal Society, *Philosophical Transactions*, XXXVII (1732), p. 266.

(10) *Ibid.*, pp. 261-73.

As a meteorologist, ROBIE was not only an observer, but also a theorist in the sense that he sought in the works of others for explanations for the phenomena observed. Some of the results of these searchings he published in his *Almanack* for the year 1716. First there is an exposition on " Why it is Hottest in Summer when the Sun is vastly furthest from us; And Coldest in Winter when it is nearest " which he obtained from the footnote on pages 87-89 of DERHAM's *Astro-Theology* (11). Following this there is a discussion on thunder, lightning, and hail. The basis of his explanations of these phenomena he found in COTTON MATHER's *Christian Philosopher*, and in JOHN WALLIS' paper in the Transactions of the Royal Society (12). MATHER suggested that lightning is the result of the vaporization and burning of combustible liquids in the earth, making an analogy to mine explosions. A similar analogy, that of the explosion of gunpowder, suggested to WALLIS that lightning was the burning of nitrous and sulphureous vapours in the air, the nitrous vapour causing the great explosion and the sulphureous the smell. It is interesting to note that FRANKLIN's hypothesis, which he was subsequently able to prove correct, was arrived at by exactly the same method : namely, by analogy to an electric spark.

In ROBIE's work there is, of course, no conception of lightning as an electrical phenomenon, and the two hypotheses which ROBIE selected for an explanation of thunder and lightning were later shown to be incorrect. Nevertheless, he did choose hypotheses which had been obtained by the method of analogy; and he did show, in recognizing the validity of the hypotheses put forth by MATHER and WALLIS, a thorough understanding of the scientific method.

Springs, wind and rain also attracted ROBIE's curiosity. In his Commonplace Book there is a group of notes entitled " Of Clouds and Rain " which he took from Dr. GARDEN's paper (13).

(11) WILLIAM DERHAM, *Astro-Theology* (London: Printed for W. INNYS, 1715).

(12) JOHN WALLIS, " A Letter... concerning the Generation of Hail, and of Thunder and Lightning...," Royal Society, *Philosophical Transactions*, XIX (1695-97), pp. 653-58.

(13) GEORGE GARDEN, " The Changes of Weather from the Alterations of the Gravity of the Atmosphere," Royal Society, *Philosophical Transactions... Abridged* (London: Printed for ROBERT KNAPLOCK, 1716), II, pp. 118-21. These particular notes appear to have been taken from the abridged edition rather than the original.

He also made references to " Dr. WALLIS in Phylosophical Transactions Abr p 123," a criticism of GARDEN's work. Still another source of ROBIE's meteorological information was THOMAS ROBINSON's *Essay towards a Natural History of Westmorland and Cumberland* (14) from which he obtained material for notes on the subject " Of the Original of Wind and Rain."

ROBIE was much more interested in astronomy than meteorology, however, to judge from his correspondence, reading, and publications. He had been instructed in the Copernican system at Harvard, where it had been taught since 1659 and probably earlier (15), but more important still is the fact that he was a Newtonian. Although there is no evidence that he ever read the *Principia*, he learned of the new mechanics from several of the early sources. The *Transactions* of the Royal Society, which were available to ROBIE, contained many Newtonian papers, and he considered at least one of them important enough to record in his Commonplace Book. His copy of HALLEY's article on tides (16) is so letter-perfect and the reproductions of the illustrations so photographic that it is certain that ROBIE used the original printing and not one of the later abridgments. As is evident from the title, this paper is an abstract of NEWTON's work. Another source of ROBIE's knowledge of Newtonian dynamics was WILLIAM DERHAM's *Astro-Theology* mentioned above. Chapter III in the sixth Book of this work is entitled " Of the Power and Usefulness of Gravity to retain the Planets within their Orbits," and DERHAM's discussion of this subject " ... is grounded upon the supposition of the truth of the Newtonian Philosophy..." (17).

ROBIE's Commonplace Book reveals that he had used at least two other important Newtonian works. The first of these two books was JOHN KEILL's *Introductio ad Veram Physicam*, on which he took extensive notes. An ardent Newtonian, KEILL was

(14) (London : Printed by J. L. for W. FREEMAN, 1709).

(15) S. E. MORISON, *Harvard College in the Seventeenth Century* (Cambridge, Mass. : Harvard University Press, 1936), I, p. 216.

(16) EDMUND HALLEY, " The true Theory of the Tides, extracted from that admired Treatise of Mr. ISAAC NEWTON, Intituled, Philosophiae Naturalis Principia Mathematica," Royal Society, *Philosophical Transactions*, XIX (1695-97), pp. 445-57.

(17) *Op. cit.*, p. 145.

NEWTON's chief protagonist in the controversy with LEIBNITZ and other continental mathematicians over the priority of the invention of the calculus. His book, consiting of the lectures which he gave as Sedleian Professor at Oxford, " ... was generally welcomed as an excellent introduction to the 'Principia' of NEWTON" (18). The second indication in ROBIE's notes of his Newtonian thinking is his praise of the Tables in WILLIAM WHISTON's *Praelectiones Astronomicae*, which he later used to calculate an eclipse of the sun (19). WHISTON was one of the first in the early group of men who popularized NEWTON's theories.

He not only read about the new theories but also accepted and used them. The best illustration of ROBIE's knowledge and use of NEWTON's work is in his *Almanack* for 1720. At the end of this little pamphlet there is an appendix which contains " ... an account of the Solar System, according to COPERNICUS and the Modern Astronomers." In this appendix there are tables of " The Middle Distances of the Planets from the Sun " and of " The Diameter of the Sun and Planets with the Moon " which, ROBIE stated, had been " ... calculated from the latest Observations by Sir ISAAC NEWTON's Rules." Here is definite confirmation of an early Colonial adherent to the new mechanics.

Harvard supplied ROBIE with several pieces of apparatus for astronomical observing. The largest instrument available was an 8-foot telescope, which the college had acquired in 1712. Later, in 1722, THOMAS HOLLIS presented the college with a 24-foot refractor, which ROBIE used to make his observations of the eclipse of the sun in the same year. The evidence of ROBIE's early activity as an observer is found in his Commonplace Book and in his papers, of which extracts were published by WILLIAM DERHAM in the *Transactions* of the Royal Society (20). The Commonplace Book contains copies of parts of three letters from WRIGHT and one from DERHAM which tell of his active interest in astronomical phenomena. Sometime after March, 1715/16

(18) R. E. ANDERSEN, " Keill, John (1671-1721)," *The Dictionary of National Biography* (London: Oxford University Press, H. MILFORD, [1921-22]), X, p. 1198.

(19) THOMAS ROBIE, " For the Entertainment of the Country and the Promoting of Knowledge," *Boston News-Letter*, Nov. 5, 1722, and the *New-England Courant*, No. 67, Nov. 12, 1722.

(20) *Op. cit.*, pp. 269-73.

—the latest date of observations mentioned in the text—ROBIE received his first letter from MATTHEW WRIGHT. In this letter WRIGHT mentioned ROBIE's observations of eclipses on December 6, 1713 and August 20, 1714, and of an immersion of the first satellite of Jupiter. He stated that he had received these observations from DERHAM, to whom ROBIE had apparently sent them; and requested ROBIE to send him any dependable observations, which he would communicate to the Royal Society. He also included his own observations of the immersion and emersion of Jupiter's first satellite. In August, 1717, WRIGHT sent more observations of the eclipses of Jupiter's satellites, and shortly afterward ROBIE received some observations of eclipses which were sent to him by DERHAM. ROBIE was apparently very much interested in Jupiter's satellites, for he had written to WRIGHT on July 23, 1717, asking for tables of the eclipses of the three outermost satellites; these were, of course, the second, third and fourth. Excusing his tardiness, WRIGHT replied to this request in a letter dated " Crew, Sept. 30, 1719," and informed ROBIE that there were no tables of the desired material in print.

Further records of ROBIE's scientific work appear in the *Philosophical Transactions* (21). The first is a note on an earthquake at Salem on February 12, 1715/16. This is followed by notes on observations of the immersions and emersions of the satellites of Jupiter, solar eclipses, southings of the moon, meteors, the aurora borealis, the zodiacal light, an eclipse of the moon on June 28, 1721, and an abstract of his observations on the solar eclipse in 1722.

In the same period ROBIE also published two scientific papers. The first was an eight-page pamphlet (22) in which he described, and gave the natural causes of a meteor which appeared on the night of December 11, 1719. The description which ROBIE gave is not that of a meteor, however, but of a remarkable display of the aurora borealis. He himself suggested that it might have been an aurora, and it undoubtedly was, for there is at least one other record of an observation of the northern lights on this

(21) *Ibid.*

(22) THOMAS ROBIE, *A Letter to a Certain Gentleman Desiring a Particular Account May be Given of a Wonderful Meteor that Appeared in New-England, on Decemb.* 11, 1719, *In the Evening* (Boston : J. FRANKLIN for D. HENCHMAN, 1719).

evening (23). Being a true man of science, ROBIE refused to interpret the dreadful red display in the heavens in terms of divinity or philosophy; instead he explained it by the use of WALLIS' nitrous-sulphureous theories mentioned above. Since the preceding days had been hot and sultry, he concluded that the combustible particles had vaporized from the earth, igniting in some manner to cause the fiery exhibition. He regarded the phenomenon as a perfectly natural one, and expressed an abhorrence toward any prognostications based upon it. Although his scientific interpretation of the aurora borealis would be untenable to-day, his attitude toward it was indeed healthy.

ROBIE's second publication (24), a paper which DERHAM communicated to the Royal Society for him, reveals that he had a considerable knowledge of chemistry. It contains a description and explanation of an abnormally high yield of alkali salt which he obtained when he burned the rotten wood of a white oak tree.

Sometime in early May, 1720, ROBIE wrote a letter to President LEVERETT of Harvard requesting " ... the Income of Mr. THOMAS BRATTLE's gift of £ 200 towards the Maintenance of a Master of Arts skill'd in the Mathematicks &c..." (25). On May 24, 1720, President LEVERETT referred ROBIE's request to the Harvard Corporation, who indicated that they would like to see an account of the mathematical performances mentioned by ROBIE in his letter. Further action on this matter was delayed for nearly a year. His account was evidently satisfactory, however, for on May 2, 1721, " Mr. ROBIE having laid before the Corporation Copys of his Mathematical Observations, particularly Such as he has transmitted to the Learned in England, [the Corporation] Voted, That £ 24 the Income of Mr. THOMAS BRATTLE's Donation for two Years be paid to Mr. ROBIE " (26).

Continuing the independent work for which he had been thus rewarded, ROBIE further pursued his astronomical observations.

(23) JOSEPH LOVERING, *On the Periodicity of the Aurora Borealis* ([Cambridge and Boston] : 1868), [From the Memoirs of the American Academy, New Series, X], p. 245.

(24) THOMAS ROBIE, " An Account of a large Quantity of Alcalious Salt produced by burning rotten Wood," Royal Society, *Philosophical Transactions*, XXXI (1720-21), pp. 121-24.

(25) *Harvard College Records*, II, p. 449.

(26) *Ibid.*, II, p. 458.

His next major investigation in this field was occasioned by the annular solar eclipse on November 27, 1722, which excited him considerably. He had a wonderful new 24-foot telescope with which to make his observations, and he made elaborate arrangements for obtaining as much data as possible. He prepared the people of the Colony for the coming phenomenon by his article " For the Entertainment of the Country and the Promoting of Knowledge " (27), which appeared in at least two Boston newspapers. In this article he described the coming eclipse, the extent of which he had calculated would be eleven and a half digits; he also requested that the " curious " make observations, particularly of the corona, and send them to him. When the sun rose on the morning of November 27th, ROBIE was at the 24-foot telescope and his assistant DANFORTH at the older 8-foot instrument. The event had been so well publicized that most of the members of the college had apparently ascended to the roofs of the buildings, from which vantage point they were able to see the eclipsed sun before ROBIE (28). Unfortunately the sun immediately clouded over, so that ROBIE was unable by observing it at its greatest phase to verify his calculations. When he next saw it an hour later it was eclipsed about six digits, and the moon passed off the face of the sun at 9:25 A.M. Apparently in response to his advertisement, he received several observations from others which enabled him partially to confirm his own calculations of the path of the shadow.

ROBIE resigned from his offices of Tutor and Fellow at Harvard on February 5, 1722/23 (29), and shortly afterward moved to Salem, where he began to practise medicine. SHIPTON points out that the probable reason for this change was ROBIE's marriage to MEHITABLE SEWALL of Salem on January 17, 1722/23 (30).

(27) *Op. cit.*

(28) THOMAS ROBIE, " Part of a Letter... Concerning the Effects of Inoculation; The Eclipse of the Sun in November 1722; And the Venom of Spiders," Royal Society, *Philosophical Transactions*, XXXIII (1724), pp. 67-70. The same notes on this eclipse also appear in an unaddressed letter from ROBIE which was probably sent to Governor WILLIAM BURNET at New York on November 9, 1723. This letter is now in the Cadwallader Colden Papers : New York Historical Society, *Cadwallader Colden Papers* (Collections, 1917, New York : Printed for the Society, 1918), I, pp. 157-59.

(29) *Harvard College Records*, II, p. 479 and p. 483.

(30) *Op. cit.*, p. 453.

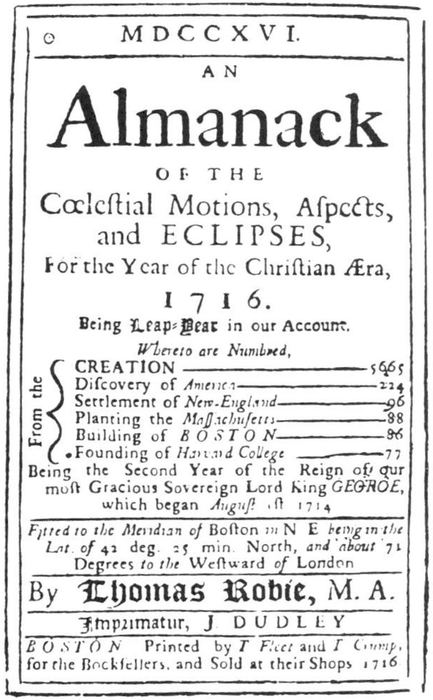

MDCCXVI.

AN

Almanack

OF THE

Cœleſtial Motions, Aſpects, and ECLIPSES,

For the Year of the Chriſtian Æra,

1716.

Being Leap-Year in our Account.

Whereto are Numbred,

From the		
	CREATION	5665
	Diſcovery of *America*	224
	Settlement of *New-England*	96
	Planting the *Maſſachuſetts*	88
	Building of *BOSTON*	86
	Founding of *Harvard College*	77

Being the Second Year of the Reign of our moſt Gracious Sovereign Lord King *GEORGE*, which began *August* 1ſt 1714

Fitted to the Meridian of Boſton *in* N E *being in the Lat. of* 42 deg. 25 min. North, *and about* 71 Degrees *to the* Weſtward *of* London

By **Thomas Robie**, M. A.

Imprimatur, J DUDLEY

BOSTON Printed by *T Fleet* and *T Crump*, for the Bookſellers, and Sold at their Shops 1716

FIG. 1

Title-page of ROBIE's *Almanack* for 1716 (Courtesy of the Harvard College Library).

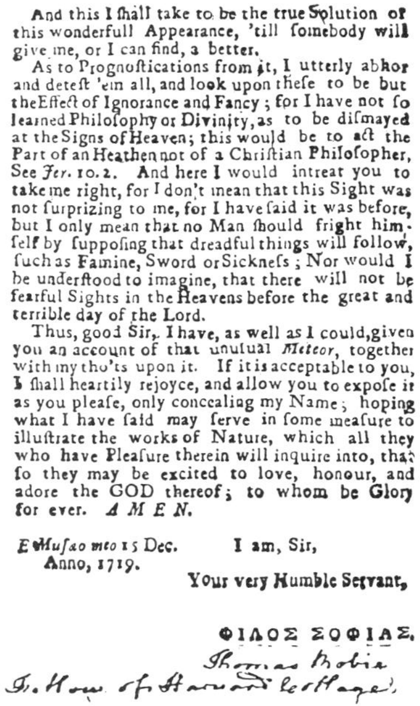

And this I ſhall take to be the true Solution of this wonderfull Appearance, 'till ſomebody will give me, or I can find, a better.

As to Prognoſtications from it, I utterly abhor and deteſt 'em all, and look upon theſe to be but the Effect of Ignorance and Fancy; for I have not ſo learned Philoſophy or Divinity, as to be diſmayed at the Signs of Heaven; this would be to act the Part of an Heathen not of a Chriſtian Philoſopher, See *Jer.* 10.2. And here I would intreat you to take me right, for I don't mean that this Sight was not ſurprizing to me, for I have ſaid it was before, but I only mean that no Man ſhould fright himſelf by ſuppoſing that dreadful things will follow, ſuch as Famine, Sword or Sickneſs; Nor would I be underſtood to imagine, that there will not be fearful Sights in the Heavens before the great and terrible day of the Lord.

Thus, good Sir, I have, as well as I could, given you an account of that unuſual *Meteor*, together with my tho'ts upon it. If it is acceptable to you, I ſhall heartily rejoyce, and allow you to expoſe it as you pleaſe, only concealing my Name; hoping what I have ſaid may ſerve in ſome meaſure to illuſtrate the works of Nature, which all they who have Pleaſure therein will inquire into, that ſo they may be excited to love, honour, and adore the GOD thereof; to whom be Glory for ever. *AMEN.*

E Muſæo meo 15 Dec. Anno, 1719.

I am, Sir,

Your very Humble Servant,

ΦΙΛΟΣ ΣΟΦΙΑΣ.

Thomas Robie

To Mr. How of Harvard Colledge.

FIG. 2

Last page of ROBIE's pamphlet on the so-called meteor (1719). Note his adherence to scientific explanations of natural phenomena and his opposition to superstition (Courtesy of the Boston Public Library)

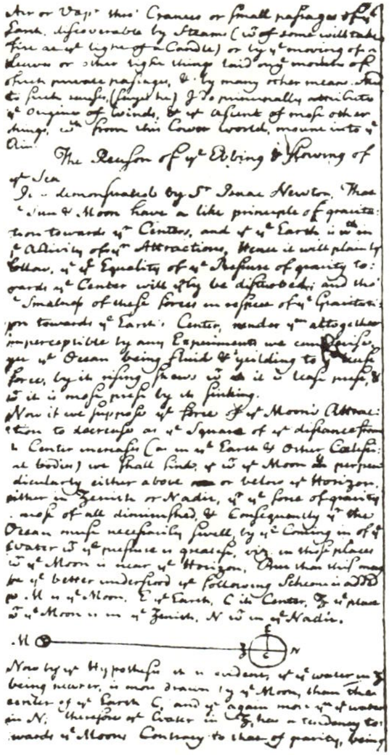

Fig. 3

Part of Robie's notes taken from Halley's Newtonian article on Tides (Courtesy of the Massachusetts Historical Society). Observe how accurately Robie copied the original shown in Figure 4.

(450)

force of Gravity towards the Center, would continue in a perfect stagnation, always at the same height, without ever Ebbing or Flowing; but it being here demonstrated, that the Sun and Moon have a like Principle of Gravitation towards their Centers, and that the Earth is within the Activity of their Attractions, it will plainly follow, that the Equality of the pressure of Gravity towards the Center will thereby be disturbed; and tho' the smallness of these Forces, in respect of the Gravitation towards the Earths Center, renders them altogether imperceptible by any Experiments we can devise, yet the Ocean being fluid and yielding to the least force, by its rising shews where it is less prest, and where it is more prest by its sinking.

Now if we suppose the force of the Moons attraction to decrease as the Square of the Distance from its Center increases (as in the Earth and other Celestial Bodies) we shall find, that where the Moon is perpendicularly either above or below the Horizon, either in Zenith or Nadir, there the force of Gravity is most of all diminished, and consequently that there the Ocean must necessarily swell by the coming in of the Water from those parts where the Pressure is greatest, *viz.* in those places where the Moon is near the Horizon: but that this may be the better understood, I thought it needful to add the following Figure, where *M* is the Moon, *E* the Earth, *C* its Centre, and *Z* the place where the Moon is in the Zenith, *N* where in the Nadir.

Now

Fig. 4

A page of Halley's paper on Tides (Op. cit.), which he based on Newton's *Principia*. Compare with Robie's notes (Fig. 3)

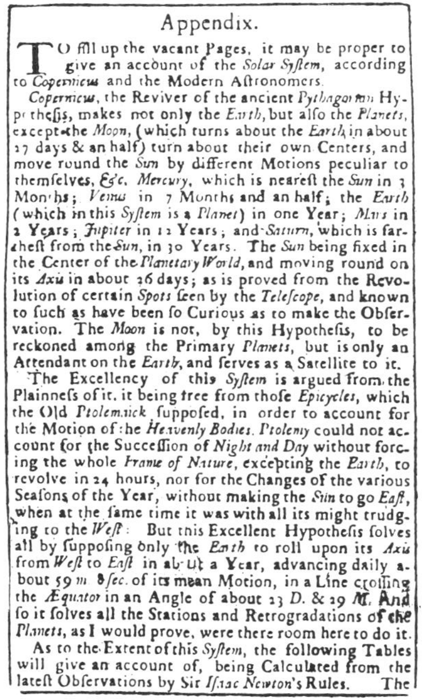

Appendix.

TO fill up the vacant Pages, it may be proper to give an account of the *Solar Syſtem*, according to *Copernicus* and the Modern Aſtronomers.

Copernicus, the Reviver of the ancient *Pythagorean* Hypotheſis, makes not only the *Earth*, but alſo the *Planets*, except the *Moon*, (which turns about the *Earth* in about 27 days & an half) turn about their own Centers, and move round the *Sun* by different Motions peculiar to themſelves, *&c. Mercury*, which is neareſt the *Sun* in 3 Months; *Venus* in 7 Months and an half; the *Earth* (which in this *Syſtem* is a *Planet*) in one Year; *Mars* in 2 Years; *Jupiter* in 12 Years; and *Saturn*, which is fartheſt from the *Sun*, in 30 Years. The *Sun* being fixed in the Center of the *Planetary World*, and moving round on its *Axis* in about 26 days; as is proved from the Revolution of certain *Spots* ſeen by the *Teleſcope*, and known to ſuch as have been ſo Curious as to make the Obſervation. The *Moon* is not, by this Hypotheſis, to be reckoned among the Primary *Planets*, but is only an Attendant on the *Earth*, and ſerves as a Satellite to it.

The Excellency of this *Syſtem* is argued from the Plainneſs of it, it being free from thoſe *Epicycles*, which the Old *Ptolemaick* ſuppoſed, in order to account for the Motion of the *Heavenly Bodies*. *Ptolemy* could not account for the Succeſſion of *Night and Day* without forcing the whole *Frame of Nature*, excepting the *Earth*, to revolve in 24 hours, nor for the Changes of the various Seaſons of the Year, without making the *Sun* to go *Eaſt*, when at the ſame time it was with all its might trudging to the *Weſt*: But this Excellent Hypotheſis ſolves all by ſuppoſing only the *Earth* to roll upon its *Axis* from *Weſt* to *Eaſt* in about a Year, advancing daily about 59 *m.* 8 *ſec.* of its mean Motion, in a Line croſſing the *Æquator* in an Angle of about 23 *D.* & 29 *M.* And ſo it ſolves all the Stations and Retrogradations of the *Planets*, as I would prove, were there room here to do it.

As to the Extent of this *Syſtem*, the following Tables will give an account of, being Calculated from the lateſt Obſervations by Sir *Iſaac Newton*'s Rules. The

FIG. 5

A page from ROBIE's *Almanack* for 1720 (Courtesy of the Massachusetts Historical Society). Note Robie's use of Newtonian mechanics as revealed in the last sentence

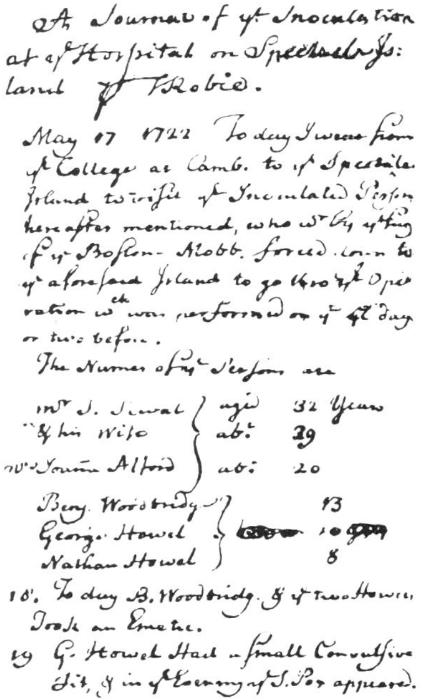

A Journal of ye Inoculation at ye Hospital on Spectacle Island by T Robie.

May 17 1722 To day I went from ye College at Camb. to ye Spectacle Island to visit ye Inoculated Persons hereafter mentioned, who wr by ye fury of ye Boston Mobb forced down to ye aforesaid Island to go thro ye Operation wch was performed on ye 4th day or two before.

The Names of ye Persons are

Mr J. Sewal aged 32 Years
& his Wife ab: 29
Mrs Joanna Alford ab: 20

Benj. Woodbridg 13
George Howel 10
Nathan Howel 8

18. To day B. Woodbridg & ye two Howels Took an Emetic.

19 G. Howel Had a small Convulsive Fit, & in ye Evening ye S. Pox appeared.

FIG. 6

First page of the MS Journal which ROBIE kept while he was taking care of those inoculated for smallpox and confined to Spectacle Island (facsimile of the original entered in RECOMPENSE WADSWORTH's Commonplace Book at the Massachusetts Historical Society)

It was not without regret that the Harvard authorities let him go, and President LEVERETT recorded the following in his MS " Book Relating to College Affairs " (31) :

> It ought to be Remembered, That Mr. ROBIE was no small honour to Harvard College, by his Mathematical Performances and by his Correspondence thereupon with Mr. DERHAM & other learned persons in those Studies abroad.

In addition to his meteorological and astronomical studies at Harvard, ROBIE had also acquired by his own diligence a comprehensive knowledge of medicine. It was undoubtedly by teaching himself that he had attained the knowledge which made it possible for him to practise medicine later, as there were no Medical Schools in the Colonies at this time. The usual procedure was to go to some European university, such as Edinburgh, or to serve an apprenticeship to a practising physician. Since only a few young men could afford the expense of a European schooling, most of the Colonial doctors were trained by the apprentice method. It is certain that ROBIE did not obtain any medical instruction abroad, but neither is there evidence that he served as an apprentice to any of the local practitioners.

The one clue to the source of ROBIE's medical education is his Commonplace Book the major part of which consists of notes on medical subjects. Taken from widely divergent sources, some of these notes are extracts from such authorities as THOMAS SYDENHAM (1624-1689), THOMAS WILLIS (1621-1675), and ARCHIBALD PITCAIRNE (1652-1713); some of them are " Medicinal Receipts " from the local townspeople, which were made up of the conglomerations typical of remedies of the period. That this Commonplace Book must have played an important function in ROBIE's practise of medicine may be inferred from the following headings which are representative of ROBIE's notes: " A Catalogue of the Diseases treated of in the Following Pages," " Medicinal Receipts from a mss," " Ad Luem Veneream inveteratam Curandam Saepius Prob." (with the text also in Latin), and " A Catalogue of Diseases and the medicines proper for them."

---

(31) In the Harvard College Library, p. 252.

ROBIE achieved at least a small right to medical immortality for his work in connection with the smallpox epidemic which broke out in Boston in 1721. Upon the eruption of this dreaded disease, the Reverend COTTON MATHER requested the Boston doctors to use the new treatment of inoculation which had been described in the *Transactions* of the Royal Society (32). All refused but one—ZABDIEL BOYLSTON, who inoculated about 240 people in spite of continuous opposition. Later two other men, Dr. THOMPSON of Roxbury and " Dr. ROBY " of Cambridge, joined BOYLSTON. According to BOYLSTON, THOMPSON performed about twenty-eight operations and ROBIE eleven (33), so that the total number of inoculations was approximately 280. Of this number only six died, giving a mortality of slightly more than 2.1 per cent, in contrast to the 14.9 per cent mortality of all those who contracted the disease. The value of the work done by the inoculators may be more fully appreciated by comparing the inoculation mortality of 2.1 per cent with the 8.5 per cent mortality of the total population.

Notwithstanding this commendable record, the opponents to variolation (and they were legion) were extremely antagonistic, and placed every conceivable obstacle in BOYLSTON's path. Finally after BOYLSTON had inoculated six people on May 11, 1722, the Boston Selectmen forced the inoculated to go down to the isolation hospital on Spectacle Island in Boston Harbor (34). This action was taken on either the 15th or 16th of May, and on the 17th ROBIE followed the inoculated down to the hospital to take care of them. While he was there he kept a journal, which he later entered in RECOMPENSE WADSWORTH's MS Commonplace Book (35) with other case histories. The following is a free transcription of the record of his two-weeks' stay on the island (36) :

---

(32) R. H. FITZ, *Zabdiel Boylston, Inoculator, and the Epidemic of Smallpox in Boston in* 1721 [Boston (?) : 1911], [From *The Johns Hopkins Hospital Bulletin*, XXII, No. 247, September, 1911], pp. 7-10.

(33) ZABDIEL BOYLSTON, *An Historical Account of the Smallpox Inoculated in New England* (2[d] ed., cor., Boston : Reprinted for S. GERRISH and T. HANCOCK, 1730), p. 32.

(34) FITZ, *op. cit.*, pp. 24-25.

(35) At the Massachusetts Historical Society.

(36) Printed with the permission of the Massachusetts Historical Society.

A JOURNAL OF THE INOCULATION AT THE HOSPITAL ON SPECTACLE ISLAND

BY

T. ROBIE

May 17, 1722. To-day I went from the College at Cambridge to the Spectacle Island to visit the inoculated persons hereafter mentioned, who were by the fury of the Boston mob, forced down to the aforesaid island to go through the operation which was performed on them a day or two before.

The names of the persons are :

Mr. S. SEWALL	Aged	32	years
and his wife	About	29	»
Mrs. JOANNA ALFORD	About	20	»
BENJ. WOODBRIDGE		13	»
GEORGE HOWEL		10	»
NATHAN HOWEL		8	»

May 18. To-day B. WOODBRIDGE and the two HOWELS took an emetic.

May 19. G. HOWEL had a small convulsive fit, and in the evening the smallpox appeared.

N. HOWEL had a flux which lasted two or three days and then the smallpox came out well. These three boys had but a few pustles and very distinct.

B. WOODBRIDGE by his carelessness got cold and so had a bad sore throat and difficulty of breathing. But by the application of blisters and other means he soon got well.

The boys, as soon as the eruption was complete, played about the house and complained no more except B. WOODBRIDGE as above.

May 20. Mr. SAM SEWALL and wife were vomited on this day and the pox appeared soon after. They both were bled on the night of May 22nd and were very faint and restless.

Mrs. SEWALL had her monthly purgation though it was not the usual time.

They both were very full and very sore, their incisions ran very much, and were very offensive. They each had an anodyne which heated Mr. SEWALL too much but was very agreeable to him, causing him to rest very well, and he continued the use of it till the pox were all turned. Mr. SEWALL spit something considerable.

May 26. This morning Mr. SEWALL was griped pretty much and had a stool very fetid about 4 in the morning, another at 10, and three at 2 P.M., and fearing he would fall into a flux which he is prone to, I gave him medicines to stop it which succeeded according to my desire and to-day we all allowed the pox to be turning.

Note : I never observed a flux hurt anybody who was bound while the smallpox was coming out and ripening, upon the turn, provided it did not exceed.

Mrs. JOANNA ALFORD was vomited on Tuesday, May 22nd, was faint and tired after it. A few pustles were seen in the skin the day before, on the 23rd they came out full. She was very sore and very restless and very full and fatter than is usual in inoculation, which I attribute to the concern of mind they were under, to the hotness of the season and their keeping them too hot before their decumbiture for fear of taking cold which I could not prevent by all I could say.

Mrs. ALFORD's fever kept up pretty long though not high.

May 24. This night she was very restless and on the 25th they began to turn. She took an anodyne during this night and slept well after it.

Note : Her menses came down also though it was not the time for nearly 9 or 10 days.

N. B. : That I never looked after a woman in the smallpox but she had her menses, let it be the time or not.

June 1. Since the 25th they have been growing better and I left them this day and returned to the College the next. And this is all I can remember about the inoculated at Spectacle Island.

Since my return I hear that Mr. SEWALL was troubled with the strangury etc. but is now well. He came to Boston again in July.

Entered from a loose piece of paper August 14, 1722.

Smallpox took a dreadful toll in the eighteenth century, its repeated epidemics being justly feared by all who had not already had the malady. The Colonies frequently suffered recurrences of this terrifying affection, and it is significant that the first medical publication to be printed in the British Colonies was THOMAS THACHER's *A Brief Rule to Guide the Common People of New-England how to order themselves and theirs in the Small Pocks, or Measles* (Boston : Printed and sold by JOHN FOSTER, 1677) (37). ROBIE understood the importance of being competent in fighting smallpox long before he heard of variolation. He devotes more space in his medical notes to this malady than to any other disease except fevers. His information about smallpox came from several

(37) See complete facsimile of this broadside in *Isis*, XXIII (1935), opposite p. 398, with commentary by HENRY R. VIETS.

sources, including SYDENHAM and ARCHIBALD PITCAIRNE's *Method of Curing the Small Pox written in the year* 1714 *for the use of the Noble and Honourable Family of March.* There is no mention of variolation, however, except in the case histories of his Cambridge smallpox inoculations which are in RECOMPENSE WADSWORTH's Commonplace Book. Since the vast majority of physicians and laymen were bitterly opposed to variolation, ROBIE should be justly given praise for his recognition and use of the new treatment. It was a small but intelligent and courageous part which he played in the first extensive use of inoculation in the Western World.

ROBIE was practising medicine in Cambridge prior to his resignation from Harvard in 1723, and he must have been exceedingly industrious to be able to give medical treatment in addition to performing his scientific investigations and discharging his responsibilities for instruction at the college. A manuscript bill (38) for medical services rendered to President LEVERETT and his household indicates that he was giving treatment before 1723. Although the bill is dated " Salem, Nov. 19, 1725," it reveals that some of the debt for medical care was incurred in 1722 and in January, 1723. ROBIE has also recorded that he was called in consulation on the case of the spider bite in September, 1722, which he later described in the Royal Society's *Transactions* (39).

After ROBIE moved to Salem the practise of medicine became his principal work, but even then he did not give up his former scientific pursuits. Although little is known of ROBIE's life in Salem, his letter to Governor BURNET (40) shows that he continued his astronomical observations, for he described the transit of Mercury, which he had observed at Salem on October 29, 1723, with a nine-foot telescope. He had been eager to determine the presence of an atmosphere around Mercury; he concluded from his observations that there was one, which he judged to be larger in proportion than that of the earth. ROBIE told BURNET that he intended to continue his observing, but if he did so there is no record of it.

(38) In the New York Public Library.
(39) *Op. cit.*, pp. 69-70.
(40) *Op. cit.*, p. 159.

On April 15, 1725, two years after he had moved to Salem, ROBIE was elected a Fellow of the Royal Society (41). He had evidently not heard of this honor by June 7th, for in a letter of this date addressed to JAMES JURIN, the Secretary of the Society, ROBIE did not mention his election. Since he had never corresponded with JURIN before, he certainly would have acknowledged his election if he had known of it. With the exception of his will this letter, now in the archives of the Royal Society, is the latest of the ROBIE documents extant. The letter mentions his meteorological observations, which he sent to JURIN at ISAAC GREENWOOD's suggestion, and concludes with some rather acute remarks on hysteria. From June, 1725, until his death, nothing specific is known of ROBIE's scientific activity. Presumably he continued to practise medicine in Salem, where he died on August 28, 1729. A Boston newspaper's account of his death accurately described him as a "... Person truly valuable and much esteemed for his Piety Learning and extensive Usefulness, having sustained and faithfully discharged the Duties of several Publick Stations" (42).

The THOMAS ROBIE who left Harvard in 1723 was a different man from the one who had received his A. B. there in 1708. During the fifteen years which followed his graduation he underwent a truly remarkable scientific evolution. He was not exactly alone in a scientific wilderness, but there was no one who could give him instruction more advanced than that which he had received in college. Nevertheless, he had the curiosity requisite for stimulating his own development. Not only did he forge ahead without the companionship of kindred scientists, but he also worked under the handicap of a lack of sufficient apparatus. For instance, he had to make his meteorological observations without the services of either a thermometer or a barometer, because neither of these instruments was available (43).

ROBIE's mind was chiefly analytical; he displayed no propensity toward being an individual theorist. He was capable of judging the theoretical work of others, however, and his acceptance of

---

(41) Royal Society, *The Record of the Royal Society of London* (3rd. ed., London : Printed for the Royal Society at the Oxford University Press, 1912), p. 331.

(42) *New-England Weekly Journal*, Sept. 1, 1729, p. 3.

(43) DERHAM, "An Abstract of the Meteorological Diaries" pp. 266-67.

NEWTON's ideas is evidence of his openmindedness in this direction. Other examples of the refreshing scientific attitude which he evolved almost independently of his fellows are the stand which he took in the inoculation episode and his active opposition to superstition. In his pamphlet on the so-called meteor, he made this opposition very clear when he said " ... I don't mean that this Sight was not surprizing to me, for I have said it was before, but I only mean that no Man should fright himself by supposing that dreadful things will follow..." (44).

ROBIE made no important contributions to knowledge, nor did he train any important scientists. It is probable, however, that he had a considerable influence on the career of ISAAC GREENWOOD, who graduated in 1721 from Harvard where he later became the first Hollis Professor of Mathematics and Natural Philosophy. From ROBIE's letter to JURIN (45), it is positive that he was corresponding with GREENWOOD while the latter was obtaining an advanced scientific training in London, but it would be unwarranted to do much more than to place ROBIE in the stream of early science in the Colonies. ROBIE's importance lies in the fact that he added to the prestige of science, and kept it going in a society that had been forced into a strictly utilitarian way of living. He advanced scientific activity in New England, and cut down the weeds of superstition. The real significance of ROBIE as a scientist lay not only in his individual achievements but also in his position in that Harvard stream of science which flowed through THOMAS BRATTLE, THOMAS ROBIE and ISAAC GREENWOOD, and which burst forth under the aegis of Professor JOHN WINTHROP.

Cambridge, Massachusetts. FREDERICK G. KILGOUR.

---

(44) *Op. cit.*, p. 8.
(45) Supra.

Philadelphia's First Scientist James Logan

By Frederick B. Tolles *

HE was only an adopted Philadelphian, but as practitioner, patron, and promoter of science, he was one of those who made the Quaker City the scientific capital of colonial America. Philadelphians thought of him as a successful businessman, a powerful politician, a man of civic spirit; but men of science on the continent of Europe, in England, in the other American colonies, knew and respected him as a virtuoso, an adept in not one but several of the provinces of natural philosophy. He was a kind of universal man in the Renaissance tradition — statesman, writer, scientist, philosopher.

The man I am describing is not Benjamin Franklin but his elder friend and advisor James Logan. It is odd that historians — especially historians of science — know so little about Logan's career, for it illustrates and confirms all the generalizations we love to make about science in colonial America — how it tended to be generalized rather than specialized, how it came to focus increasingly in the town of Philadelphia, how its devotees transcended the limitations of provincial culture and took their places in an "Atlantic community" of scientific learning. But Logan was something more than a mere lay figure on which to drape historical generalizations. In his own right he was a scientist as original, as productive, as influential, as distinguished as any American before Franklin. As a mere dabbler in the history of science all I can do here is to direct the attention of my betters to a man who distinctly deserves their notice.

It is not the historians of science only who have neglected Logan. Though he was, all things considered, perhaps the most noteworthy man in America in the first half of the eighteenth century, he still has no adequate biography.[1] Born in the north of Ireland in 1674, he came to Pennsylvania in 1699 as Secretary to William Penn. Over the next five decades until his death in 1751, he occupied at one time or another nearly every place of honor and responsibility in the province. As the representative of the Penn family and a member of the Council he was able by his sheer political weight to control the provincial government, to make or break Governors, to humble and unseat even so popular an official as the brilliant, unscrupulous Governor Sir William Keith.

* Friends Historical Library of Swarthmore College.

[1] Wilson Armistead's *Memoirs of James Logan* (London, 1851) is a slight work, based on inadequate sources and marred by errors. Irma Jane Cooper's *Life and Public Services of James Logan* (New York, 1921) is even slighter and, though a Ph.D. dissertation, hardly more useful. Joseph E. Johnson's A Statesman of Colonial Pennsylvania: A Study of the Private Life and Public Career of James Logan to the Year 1726, an unpublished Harvard Ph.D. thesis, is solid, judicious, and massively documented, but it terminates before the period of Logan's most intensive scientific activity. The author of this article is preparing a short biography to be published by Little, Brown, and Company.

He presided over the Pennsylvania court system from 1731 to 1739 as Chief Justice, and served a term as Mayor of Philadelphia. He was the city's leading fur merchant and amassed a sizable fortune on which he retired in 1730 to his handsome country seat of Stenton. In 1736, when he was over sixty, he emerged from retirement to serve as acting Governor during a turbulent and critical period in the province's history.

He was, moreover, the most accomplished, far-seeing, widely-respected Indian diplomat of his time, and an imperial statesman who early grasped the significance of the momentous Anglo-French struggle for the control of North America.[2] In his leisure moments he was a humanist as well as a scientist. His translations of Latin classics, first printed by Franklin in Philadelphia, were reprinted in England and Scotland.[3] One of his many learned correspondents was the German classical scholar J. A. Fabricius. The critical comments on the Greek text of Pythagoras and on the Arabian versions of Euclid which he sent to Fabricius were published at Hamburg and Amsterdam.[4] One striking incident in the friendship of these two humanists shows how real the "Atlantic community" of scholarship was in the eighteenth century: Fabricius, learning that Logan wanted the rare 1538 edition of Ptolemy's *Almagest*, sent him a copy from his own library, and the Philadelphia fur merchant reciprocated with the worthiest token in his gift — a buffalo skin.

Logan's lifelong appetite for learning was first stimulated, no doubt, by his father, an M.A. of Edinburgh and a schoolmaster by profession. From his Quaker upbringing he would naturally have absorbed a special taste for science; it is no accident that botany and mathematics, his two favorite fields, were recommended by both William Penn and Robert Barclay, the two leading Quaker intellectuals of his youth, as suitable subjects of study for Friends.[5] We know what it was that definitely set him on the road of scientific learning. It was "a book of Leyborns" — probably William Leybourn's *Cursus mathematicus* — which he encountered at the age of sixteen and promptly mastered "without any manner of Instruction." [6] Throughout his life the bent of his mind was mathematical: the greater part of his scientific work was in pure mathematics, astronomy, and physics, and he even sought to ground a science of morals on mathematical principles.[7]

[2] See Joseph E. Johnson, A Quaker Imperialist's View of the British Colonies in America, 1732, *Pennsylvania Magazine of History and Biography*, 1936, *60*: 97–130.

[3] *Cato's Moral Distichs* (Philadelphia, 1735); *M. T. Cicero's Cato Major; or, His Discourse of Old Age* (Philadelphia, 1744, 1758, ca. 1809; London, 1750, 1778; Glasgow, 1751, 1758).

[4] Hermann Samuel Reimarus, *Commentarius de vita et scriptis Io. Alb. Fabricii* (Hamburg, 1737), pp. 291–95; *Epistola ad virum clarissimum Joannem Albertum Fabricium* in J. P. d'Orville, ed., *Miscellaneae observationes criticae novae in auctores veteres et recentiores* (Amsterdam, 1740), pp. 91–112.

[5] I have discussed the relationship of Quakerism and science in *Meeting House and Counting House: The Quaker Merchants of Colonial Philadelphia* (Chapel Hill, 1948), chap. ix.

[6] Autobiographical sketch in Logan Papers, Historical Society of Pennsylvania, Vol. II; reprinted in Albert Cook Myers, *The Immigration of the Irish Quakers into Pennsylvania* (Swarthmore, Penna., 1902), p. 239. All Logan MSS cited hereafter are in the Historical Society of Pennsylvania, unless otherwise indicated.

[7] Logan's ambitious treatise on "The Duties of Man Deduced from Nature" was never published and now exists only in fragments. For partial summaries see his correspondence with Thomas Story, ed. Norman Penney, *Bulletin of Friends Historical Association*, 1926, *15*: 57–66; and his *Charge Delivered from the Bench to the Grand Inquest . . .* (Philadelphia, 1736). In the Logan Papers (II: 118–20) are some revealing notes in which he likens morality to algebra and geometry. As he completed successive chapters of this work he submitted them for criticism to

From the day when he acquired this "book of Leyborns" to the day when, over seventy years old and "much fail'd in all respects," he asked his friend Benjamin Franklin to lay out £200 in the London bookshops for him, the collecting and reading of books was his passion — his "delight," he called it, and his "disease." [8] In the course of his fifty years in America he assembled and ultimately gave to the city of Philadelphia what I do not hesitate to call the finest library in the colonies. It was not the largest — William Byrd's in Virginia, Cotton Mather's in Massachusetts exceeded it in sheer size — but no colonial library surpassed it in balance and catholicity, and no colonial library equaled it in representation of great books and rare editions.[9] In particular, no collection of books in pre-Revolutionary America, public or private, bears comparison with it as a scientific library.

There were upwards of four hundred scientific and mathematical books at Stenton — considerably more than Harvard College could boast in 1735.[10] But it is the quality, not the mere size, of the library that is impressive. At a time when there was no copy of Newton's *Principia* at Harvard, Logan had three editions and most of Newton's other works besides. Indeed, there was hardly a major scientific writer, ancient or modern, who was not represented on Logan's shelves: Archimedes, Euclid, and Ptolemy in the best editions; Roger Bacon and the Arab scientists; Copernicus, Galileo, Tycho Brahe, and Kepler; Huygens, Halley, Flamsteed, and Hevelius; Gilbert and Harvey, Leeuwenhoek and Malpighi, Linnaeus and Grew, Boerhaave and Sydenham, Boyle and Hooke — all these (and I single out only the greatest names) were in the collection at Stenton.

The existence of this great library — and Logan's generosity in giving scientifically-minded young Philadelphians the run of it — was unquestionably an important precondition of the Quaker City's emergence as the major center of scientific activity in the American colonies. In the manuscript outline of Benjamin Franklin's *Autobiography*, for instance, is this suggestive notation: "Logan fond of me. His library." John Bartram, by his own acknowledgment, had his first introduction to botanical literature when Logan lent him Salmon's *Pharmacopeia Londinensis*, Culpeper's *English Physician*, and Turner's *Herbal*. There is a pretty legend that Logan "discovered" Thomas Godfrey when he entered his library one day and found the young man, an uneducated artisan hired to set windowglass at Stenton, engrossed in reading Newton's *Principia*.

---

his scientific friends in England and America — Peter Collinson, Sir Hans Sloane, Dr. Richard Mead, William Jones, Benjamin Franklin. It may not be altogether fanciful to suggest Logan as the source of the notion of "moral algebra," which Franklin developed in a letter to Joseph Priestley, 19 September 1772 (*Writings of Benjamin Franklin*, ed. A. H. Smyth [New York, 1907], V: 437–438).

[8] Logan to Franklin, 13 July 1747, Sparks MSS, Harvard College Library, XIX: opp. p. 31; Logan to Frederick Schomaker, 14 Nov. 1726, Logan Letter Book, 1717–31, p. 215; Logan to James Steel, 3 Dec. 1729, Logan Papers, I: 93.

[9] Two catalogues exist: one in MS, dated *ca.* 1740, listing approximately 1625 titles; the other the printed *Catalogus bibliothecae Loganianae* (Philadelphia, 1760), naming approximately 2600 titles. Logan's library, unlike those of Byrd, Mather, and Thomas Jefferson, is still virtually intact. The Library Company of Philadelphia, under whose care it has been since 1792, has recently undertaken to rehabilitate and recatalogue it. The numerous marginalia in Logan's scientific books would repay study.

[10] Cf. Frederick G. Kilgour, The First Century of Scientific Books in the Harvard College Library, *Harvard Library Notes*, 1939, 3: 216.

(Logan's own account in a communication to the Royal Society is probably more authentic and serves equally well to substantiate my point about the role of Logan's library: he says he first made Godfrey's acquaintance when the young man presented himself at the door and asked to borrow the *Principia.*) [11] When Logan left his library to the city, he provided that it should be open to all "who have been educated in reading and writing, and more especially those who have any knowledge of the Latin tongue, or who study any of the mathematical sciences or medicine."

The pattern of Logan's book-buying confirms my observation that the characteristic bent of his mind was mathematical. One of his earliest important purchases was the first edition of Newton's *Principia mathematica.* He acquired it in April, 1708. The claim has been made that this was the first copy in the American colonies.[12] This is the kind of claim which can never really be substantiated, and it would be of dubious significance if it could. The more important question, in any case, is whether or not Logan understood Newton. On this point there is little room for doubt. His marginal annotations, his cross-references and critical comments show that he studied it with both zeal and understanding. Moreover, within a few months of acquiring the *Principia* he had a copy of Charles Hayes's *Treatise of Fluxions* to help him over the hard places in Book I.[13] I dare not say that Logan was the first American to master the calculus, but would simply point out that it would be ten years before fluxions would begin to appear in the commencement theses at Harvard and twenty before the method was systematically taught there by Isaac Greenwood.

Logan's admiration for Sir Isaac was almost unbounded. "The world, as long as there remains in it any regard for science and sound knowledge," he wrote, "must ever revere that wonderful man's memory and acknowledge him the greatest genius in that way that has ever been known to this day." [14] It was a high point in Logan's life when, on a visit to London in 1724, he spent two hours at a meeting of the Royal Society and saw its venerable President "bending so much under the load of years as that with some difficulty he mounted the stairs of the Society's room." [15]

Still, he was not an uncritical hero-worshiper. When he acquired a copy of the third edition of the *Principia* in 1726, his careful eye immediately noted two alterations in the text which, as he put it, showed "the prevalence of human passions even in the greatest." Wherever possible, Newton had eliminated the name of the astronomer Flamsteed, with whom he had had a famous quarrel; he had also dropped the reference in the second Book to Leibnitz's independent

[11] *Benjamin Franklin's Autobiographical Writings,* ed. Carl Van Doren (New York, 1945), p. 211; John Bartram to Sir Hans Sloane, 23 September 1743, in William Darlington, *Memorials of John Bartram and Humphry Marshall* (Philadelphia, 1849), p. 304; *Philosophical Transactions of the Royal Society,* 1734, *38*: 441–457.

[12] Frederick E. Brasch, James Logan, a Colonial Mathematical Scholar, and the First Copy of Newton's *Principia* to Arrive in the Colonies, *Proceedings of the American Philosophical Society,* 1942–43, *86*: 6.

[13] Isaac Norris to Logan, 6 March 1707/8, *Correspondence between William Penn and James Logan,* ed. Edward Armstrong, *Memoirs of the Historical Society of Pennsylvania,* 1872, *10*: 263.

[14] Logan to William Jones, 25 July 1737, *Correspondence of Scientific Men of the Seventeenth Century,* ed. Stephen J. Rigaud (Oxford, 1841), I: 316.

[15] Logan to William Burnet, 7 Feb. 1726/7, Logan Letter Books, III A, 60–64.

discovery of the calculus.[16] Logan read Newton's later works — the *Chronology of Ancient Kingdoms Amended* and the *Observations upon the Prophecies of Daniel and the Apocalypse* with mounting impatience, regretting that so great a scientist should have wandered off into what he considered unprofitable bypaths.[17]

The pleasures of mathematics continued to charm Logan to the end of his days. In 1732, after he had retired from business, he commenced a correspondence with William Jones, F.R.S., the English mathematician and editor of Newton. Here was an intellectual peer with whom he could discuss infinite series, quadratic equations, classical problems in geometry like the duplication of the cube.[18] There were only a few men in the colonies with whom he could talk the language of higher mathematics, but he made the most of his limited opportunities. He encouraged the mathematical prodigy Thomas Godfrey to develop his remarkable powers, and the two men — the one a humble artisan, the other a provincial grandee — often had their heads together over a page of abstruse formulas and equations. When Benjamin Franklin confessed that he frequently occupied his mind during the tedious debates in the Pennsylvania Assembly by constructing "magic squares," Logan, showing him some classical examples in the works of Bernard Frénicle de Bessy and Michael Stifel, convinced him that the construction of these *difficiles nugae* was a useful exercise in sharpening the mind.[19] Mathematics was one of the interests he shared with that amateur of all the sciences Cadwallader Colden. Colden lived in Philadelphia for a number of years as a young physician, and Logan tried to persuade him to stay there, as much for his intellectual companionship as for his public usefulness. After the young man moved to New York, Logan kept up a desultory correspondence with him on matters scientific and mathematical. Many years later, at Franklin's request, he looked over his old friend's treatise on fluxions and offered suggestions for improvement.[20] He distrusted Colden's tendency to soar into metaphysics, however, and to indulge in unverifiable abstractions. "For bare speculation only," he wrote, like a true mathematician, "number and measure appear to me to be the most adequate objects on earth for the human intellect."[21]

Number and measure — these were the preoccupations of Logan's working hours as well as his leisure during most of his life. For it was his responsibility as the Penn family's principal agent to supervise the running of property lines, the laying out of new tracts of land in Pennsylvania. His skill with transit and quadrant helped him in his exhaustive studies of the controversial Maryland-Pennsylvania boundary — studies which in great measure fixed the location of

[16] *Ibid.*; Logan to Silvanus Bevan, 7 Feb. 1726/7, *ibid.*, III A, 64–67; Logan to Burnet, 10 May 1727, *ibid.*, III A, 73–76.

[17] Logan to Burnet, 2 May 1728, *ibid.*, III C, 22–23; Logan to Josiah Martin, 5 Nov. 1733, *ibid.*, IV: 345–47.

[18] This correspondence is in Rigaud, Vol. I, *passim*.

[19] Franklin to Collinson, n.d., *Writings of Franklin*, II: 456–60. The work of Frénicle was surely his treatise *Sur les quarrez magiques*, not the *Traité des triangles rectangles en nombres*, as Franklin's editor suggests.

[20] Franklin to Colden, 13 Sept. 1744, *Benjamin Franklin's Autobiographical Writings*, ed. Carl Van Doren (New York, 1945), p. 44; Franklin to Colden, 25 Oct. 1744, *Collections of the New-York Historical Society*, 1919, 52, 77; Franklin to Colden, 6 Aug. 1747, *Writings of Franklin*, II: 323–24.

[21] Logan to Jones, 16 Oct. 1738, Rigaud, I: 335–41.

the definitive line which Mason and Dixon would eventually run after 1763.[22]

Personal contacts with John Flamsteed, the Astronomer Royal, stimulated his interest in astronomical observations. He visited Flamsteed at Greenwich in 1710 and later referred to him as "my good friend (while living)." [23] Early in 1717 he ordered from England a telescope of twenty-five to thirty-foot focus and mounted it in the back yard of his Philadelphia house. There he and Colden took observations in an effort to determine the annual parallax in the stars resulting from the earth's motion about the sun.[24] While he was in England in 1723, he managed to obtain from the bookseller Innys a set of sheets of Edmund Halley's unpublished astronomical tables. He brought them back to Pennsylvania and proceeded with their help to analyze the moon's orbital motion. In the course of his studies he found a number of errors in Halley's computations.[25] Years later, in 1744, when Logan was nearly seventy, a Philadelphia Quaker wrote that "the old gentleman and Tom Godfrey are very busy in inspecting into a comet that has appeared for three weeks past . . . and the public expect their opinion of it in print from some hints given." [26]

It was as young Thomas Godfrey's sponsor in the controversy with John Hadley over the invention of the reflecting quadrant that Logan first ventured to address the Royal Society. Godfrey was a member of Franklin's circle, a glazier by trade and a self-taught mathematician. Franklin found him a disagreeable companion who "expected unusual precision in everything said, or was forever denying or distinguishing upon trifles to the disturbance of all conversation." [27] But Logan — perhaps because the young man's habit of thought and speech was not unlike his own — developed a real fondness for him. One day in February, 1730/31, Godfrey showed him an improvement he had devised in the mariner's quadrant — an arrangement of mirrors by which the images of two stars could be brought to coincide. This instrument made it possible to find latitudes readily, even on the bouncing deck of a small vessel in a rough sea.

After a long and unaccountable delay, Logan wrote to Edmund Halley, the second Astronomer Royal of his acquaintance, describing his protégé's invention. The letter was dated 25 May 1732.[28] Just a year earlier — though neither Logan nor Godfrey knew it — John Hadley had described to the Royal Society a substantially similar instrument. When Logan heard of Hadley's claim, he immediately scented foul play. Edmund Halley's silence only increased his misgivings. To his suspicious mind circumstances formed a pattern which could only indicate skulduggery. It appeared that Captain Cox, who had used Godfrey's instrument on its first trial in the West Indies in January 1730/31, had indiscreetly blurted out the secret of its design one evening in a

[22] See, for example, The Claims of the Proprietors of Maryland and Pennsylvania Stated (1715), *Maryland Archives*, VII: 72 et seq.

[23] Logan Papers, I: 93.

[24] Logan to Joseph Williamson, 1 March 1716/17, Logan Letter Books, II: 10–12; also later letters, *ibid.*, 38, 40, 42.

[25] Logan's MS *Calculus motus lunae ex his tabulis Halleianis* is in the Loganian Library, Library Company of Philadelphia.

[26] Richard Hockley, quoted in Charles P. Keith, *The Provincial Councillors of Pennsylvania* (Philadelphia, 1883), p. 13. Logan's observations of the comet of 1744 are in the Logan Papers, II:13; I have not found them in print.

[27] *The Autobiography of Benjamin Franklin*, ed. Max Farrand (Berkeley, 1949), p. 74.

[28] *Phil. Trans.*, 1734, *38*: 441–57.

tavern in Kingston, Jamaica. "A gentleman sent over by the crown to take a survey of the Isle of Providence" in the Bahamas had heard him. What more likely, reasoned Logan, than that the gentleman on his return to England had described the instrument to Hadley, who had made one and claimed it as his own before the Royal Society? [29]

Logan never proved his case, but he finally got a hearing before the Royal Society by enlisting the powerful support of William Jones and John Machin, the mathematicians, and Peter Collinson, the Quaker botanist. Jones stated flatly that in his opinion Godfrey deserved credit for the independent invention of the improved quadrant, and Machin, presenting the affidavits which Logan had assembled on Godfrey's behalf, observed that if they were not valid evidence, "we must believe that all the people in Pennsylvania combined to impose on the Society, which no reasonable man will do." [30]

Mr. Brasch has made the plausible suggestion that it was because of his staunch advocacy of Godfrey's claim that Logan never was made a Fellow of the Royal Society.[31] Be that as it may, Logan continued for five years to bombard the Society with communications. In 1735, his well-known experiments on the generation of Indian corn were read — of which more hereafter. On 12 February 1735/36, Sir Hans Sloane presented a long communication from him, typically heterogeneous and wide-ranging in subject-matter. First, Logan submitted as a gift for the Society's collections an ancient Hebrew shekel, pointing out the obvious error of those numismatists like Charles Patin who asserted that the only genuine example in existence was in the Escorial in Madrid. Next, he made some supplementary comments by way of an appendix to his earlier essay on generation. Then he offered a theory to explain the crooked appearance of lightning. Finally, he proposed an answer to the ancient problem of the "horizontal moon," i.e., why the moon appears larger at the horizon than at the zenith. For these contributions he was voted thanks, and his observa-

[29] Logan to George Barclay, 5 Nov. 1733, Logan Letter Books, IV: 363. Legend has had it that it was Hadley himself (or a relative of the same name) who supposedly saw Godfrey's instrument in the West Indies. S. P. Rigaud dismissed this story by showing that there was no one of that name in the Royal Navy in 1730. See his review of the Hadley-Godfrey controversy in the *Nautical Magazine*, 1832, I: 348–52; 1833, II: 654–62; 1834, III: 201–10; 1834, III: 339–45. But the only *documentary* evidence for the story — the letter cited above, which Rigaud never saw — does not make this supposition. Moreover, there *was* a surveying expedition in the Bahamas in 1730, under the command of one Captain Gascoyne — a fact which lends some credibility to Logan's charge. *Calendar of State Papers: Colonial . . . 1730*, p. 317. There is little point, however, in reopening the controversy. Most authorities accept Godfrey as an independent inventor. In any case, both Godfrey and Hadley had been anticipated by Newton.

[30] See letters of Edward Wright to Logan, Library Company of Philadelphia, especially 4 February 1733/4. Logan's and Godfrey's letters to the Royal Society will be found in *Phil. Trans.*, 1734, *38*: 441–57. There is a persistent legend that the Royal Society or some of its members made him a gift of £200, but, knowing the young man's weakness for strong drink, took the precaution of remitting it in household furniture instead of cash. This story can be traced back at least as far as Samuel Miller's *Brief Retrospect of the Eighteenth Century* (New York, 1803), pp. 468–80, in which it is given on the authority of Alderman Michael Hillegas of Philadelphia. There is no record at the Royal Society of any official gift to Godfrey, but some slight color is lent to the story by a casual remark of Logan's which intimates that Godfrey was not without personal failings. Logan to Jones, 25 July 1737, Rigaud, I: 316–17.

[31] *Op. cit.*, p. 11. Logan's resentment against Halley can be amply documented in the Logan Papers. See, for example, Logan to Lawrence Williams, 14 Jan. 1733/34, Letter Books, IV: 385.

tions on lightning and the moon-illusion were ordered printed in the *Philosophical Transactions*.[32]

These two contributions are perhaps less important for their own sake than for the light they throw on Logan's qualities as a scientist. The jagged appearance of lightning he attributed to refraction through the clouds, though when Franklin later showed him his papers on lightning, he demonstrated his flexibility and openness of mind by accepting the new theory at once, "notwithstanding my piece in the Transactions." [33] His explanation for the moon-illusion was one which has had general acceptance until fairly recently: he related it to the size of familiar objects close at hand — in this case the trees through which he watched the moon rise at Stenton. Though in non-scientific matters the opinions of the ancients tended to carry great weight with him, he rejected out of hand the Ptolemaic idea that this particular phenomenon was caused by refraction.[34]

His interest in optical phenomena and his passion for mathematics finally converged in two treatises which have been strangely ignored by historians of science. The first, completed in 1738 on his sixty-fifth birthday, was suggested by his study of the posthumous *Dioptrics* of Christiaan Huygens. He was full of admiration for Huygens's propositions, but was shocked "at the tediousness as well as the obscurity of the demonstrations." He set himself the task of working out the latter more simply, both by analysis and by geometrical construction.[35] Succeeding beyond his expectations, he wrote out the proofs both in Latin and English and sent them to the Royal Society. Why they were not published in the *Philosophical Transactions* is not clear, though the fact that John Hadley was now Vice-President may have had something to do with it. Peter Collinson, always the friend of American scientists, sent the manuscript to Holland, where it was published in 1739.[36]

Whether Logan's proofs were valid, whether they represented an improvement over Huygens must be left to more competent judges. As an historian I would call attention to Logan's remarkable knowledge of the history of this branch of optics as displayed in the preliminary survey which he prefixed to the English version. There he discusses — with every evidence of first-hand acquaintance — the contributions of Aristotle, Euclid, Ptolemy, Alhazen, Roger Bacon, Vitellio, Galileo, Kepler, Snel, Descartes, Barrow and Gregory

[32] The complete letter, dated 20 Sept. 1735, is in the Journal Book of the Royal Society, Burlington House, London, Vol. XV: 1734–36. The remarks on lightning (Concerning the Crooked or Angular Appearance of the Streaks or Darts of Lightning in Thunder Storms) appear, with some excisions, in *Phil. Trans.*, 1736, *39*: 240. Those on the moon-illusion (Some Thoughts on the Sun and Moon, when near the Horizon Appearing Larger than when near the Zenith) are in *Phil. Trans.*, 1736, *39*: 404–405.

[33] Logan to Franklin, n.d., *Works of Benjamin Franklin*, ed. Jared Sparks (Boston, 1840), VII: 38–39n.

[34] For a history of research on the moon-illusion from Aristotle to the twentieth century, see E. Reimann, *Die scheinbare Vergrösserung der Sonne und des Mondes am Horizont*, *Zeitschrift für Psychologie*, 1902, *30*: 1–38, in which Logan's contribution is noticed at p. 17.

[35] Logan to Jones, 16 Oct. 1738, Rigaud, I: 335–41.

[36] *Canonum pro inveniendis refractionum tum simplicium, tum in lentibus duplicium focis, demonstrationes geometricae*, published with Logan's *Experimenta et meletemata de plantarum generatione* (Leyden, 1739), pp. 15–32. The English version, entitled "The Principal Rules in Dioptrics," is in the Logan Papers, II: 115ff.

— all the great names associated with the history of dioptrics down to Huygens himself.

Logan's second and more ambitious venture into mathematical optics was a work on spherical aberration. Here too he took Huygens as a point of departure, but he showed a thorough knowledge of Newton's *Opticks* as well. His purpose was to prove that, Huygens to the contrary notwithstanding, the laws of spherical aberration could be worked out with absolute mathematical rigor. His demonstrations, he flattered himself, were superior to those of either Huygens or Newton. This treatise too he published in Leyden, and I can only account for its neglect by the fact of its excessive rarity.[37]

Despite his lifelong devotion to the mathematical sciences, Logan achieved most recognition in his own time for his work in botany. His admirably designed, carefully controlled, lucidly reported experiments with Indian corn proved conclusively that the pollen was the male element and that it was necessary for the production of viable seed. The significance of these experiments in the history of plant science has been adequately treated by Professors Zirkle and Roberts.[38] I shall content myself therefore with a few observations on their date and their subsequent influence, drawing my information from the unpublished letters in the Logan Papers and elsewhere and from an unscientific sampling of the botanical literature of the eighteenth and early nineteenth centuries.

The germinal idea came from a passage in William Wollaston's *Religion of Nature Delineated*, a work of rational theology, which Logan read with great excitement in the autumn of 1726. Here he first encountered the widespread theory of preformation and the notion of wind-pollination. From Richard Bradley's *New Improvements of Planting and Gardening* he got the idea that "all plants have their male as well as female seed." [39] Starting from these hints he conceived an experiment. He would detassel some corn plants, remove the filaments from others, and observe the effect upon the development of the kernels. He carried out the experiment in the summer of 1727. The results fully confirmed his hypothesis. He shared his discovery with Thomas Goldney, a Quaker botanist of Bristol, England, urged him to perform the experiment too — a striking instance of transatlantic scientific co-operation.[40] The next year he repeated the procedures, refining and improving his experimental technique. He waited, however, till 1735 to make his findings public in a letter to Peter Collinson, which was published in the *Philosophical Transactions* early the next year.[41]

[37] *Demonstrationes de radiorum lucis in superficies sphaericas remotius ab axe incidentium a primario foco aberrationibus* (Leyden, 1741).

[38] Conway Zirkle, *The Beginnings of Plant Hybridization* (Philadelphia, 1935), p. 70; *idem*, Some Forgotten Records of Hybridization and Sex in Plants, 1716–1739, *Journal of Heredity*, 1932, *23*: 440–42. See also H. F. Roberts, *Plant Hybridization before Mendel* (Princeton, 1929), pp. 68–70.

[39] Logan to William Burnet, 7 Nov. 1726, Logan Letter Books, III A, 59–60; Logan to William Logan, 25 Sept. 1727, *ibid.*, III A, 83–84; Logan to Burnet, 10 Jan. 1727/28, *ibid.*, III C, 5.

[40] Logan to Goldney, 20 Nov. 1727, *ibid.*, III A, 87–90; 6 Nov. 1728, *ibid.*, III C, 47. Goldney performed Logan's experiments and reported success, Logan Papers, X: 47.

[41] Some Experiments concerning the Impregnation of the Seeds of Plants, *Phil. Trans.*, 1736, *39*: 192–95. Collinson (or someone else) made interpolations in the text which, com-

The Royal Society, it must be recorded, did not take much notice. Two-thirds of the members present when extracts from his letter were read were busy "dissecting a German cabbage and looking for the small fibers in the root of an Indian turnip." So reported a friend of Logan's who attended the meeting.[42] But the significance of the experiments was not lost on the international community of botanists, which reached from the colonies in North America to the centers of learning in Russia.

One of the first scientists to take note of Logan's contribution was Linnaeus. He wrote a fulsome letter of congratulation from Amsterdam in May, 1738, a letter in which Logan found himself ranked, if we can judge from his self-deprecatory reply, among the demigods of science (*Tu me inter Naturae Mystas heroa existimare videris*). With somewhat uncharacteristic modesty Logan disparaged his own attainments as a botanist and then generously called Linnaeus' attention to another but little-known American — "a certain countryman named John Bartram" (*rusticus quidam Joannes Bertram*) who, in spite of his lack of education, had a natural genius for botanizing and from whom, granted life and leisure, great things could be expected.[43]

The fame of Logan's experiments spread. In 1739 J. F. Gronovius published them in full at Leyden under the title *Experimenta et meletemata de plantarum generatione*. Within the next twenty-five years Logan's name came to be known wherever botanists worked. The experiments which this provincial politician and fur merchant had performed in his garden on Second Street in Philadelphia were cited in learned dissertations published in Stockholm and Upsala, in Tübingen and Nuremberg.[44] The *Experimenta et meletemata* were republished in London in 1747 with an English translation by Dr. John Fothergill. Linnaeus made use of Logan's work in his prize-winning dissertation on the sexuality of plants before the Imperial Academy of Sciences at St. Petersburg in 1760.[45] And when Koelreuter came to review the researches which had led

---

plained Logan, made him "speak nonsense" at several points. See his letter dated 31 Oct. 1737, Royal Society Letter Book L, 6, p. 48.

[42] Edward Wright to Logan, March, 1734/5 (*sic*; should be 1735/6), Library Company of Philadelphia.

[43] Logan to Linnaeus, 17 Oct. 1738, Royal Linnaean Society, London. This letter would seem to be Linnaeus' first introduction to Bartram. Logan had already introduced Bartram to Linnaeus' writings. In June, 1736, he had described to his young friend a new book he had just received in printer's sheets from Peter Collinson — a system of classification which he recognized as 'altogether new' and worthy of Bartram's attention. The book was the first edition (1735) of the *Systema naturae*. Logan to Bartram, 19 June 1736, Darlington, *Memorials*, pp. 307–308. Further evidence of Linnaeus' early appreciation of Logan's experiments is seen in a letter of 7 March 1738 to Peter Collinson: "I have read with a great deal of pleasure the ingenious Mr. Logan's experiments made in North America upon the sex of plants . . . good and complete experiments have hitherto been very scarce, therefore it is to be wished that those of Mr. Logan's might be soon published as they are made with skill and accuracy." (Collinson Correspondence, p. 242, at the Royal Linnaean Society; in microfilm at the American Philosophical Society).

[44] Johann Gustaf Wahlbom, *Sponsalia plantarum* (Stockholm, 1746), also in Linnaeus *Amoenitates academicae* (Erlangen, 1787), I: 329, 345, 371, 372, 374; Johann Georg Gmelin, *Sermo academicus de novorum vegetabilium post creationes divinum exortu* (Tübingen, 1749); John Mitchell, *Dissertatio brevis de principiis botanicorum et zoologicorum* in *Acta Phys. Med. Acad. Leopoldina . . . Ephemerides*, 1748 *8*: App. 178–202 (republished Nuremberg, 1769). The original letter in the *Phil. Trans.* was reprinted in the *Physicalische Belustigungen*, ed. von Mylius (Berlin, 1757), III: 1088–1102. There are further references to the European use of Logan's work in Zirkle, *Beginnings*, pp. 161, 188.

[45] *Disquisitio de sexu plantarum* (St. Petersburg, 1760), reprinted in *Amoenitates academicae*, X: 122. Collinson sent Linnaeus a copy of the 1747 edition with the suggestion that it be read "at a meeting of your learned society." *A*

up to his own magistral work on pollination and hybridization, he paid appropriate tribute to Logan for his work on "*türkisch Korn.*"[46] Though Koelreuter's work marked the end of one chapter in botanical research and the beginning of another, this was not the end of Logan's influence. His experiments continued to be cited with respect in botanical treatises well into the nineteenth century.[47]

It is only in our own time that he has been ignored and all but forgotten. Probably this situation has been partly owing to the rarity of his writings. Even in 1875, when Julius Sachs wrote his *History of Botany*, he could locate no copy of Logan's *Experimenta et meletemata* and was obliged to rely on Koelreuter's summary.[48] But whatever the reason for it, this neglect of Logan as a scientist has been unfortunate. It is time to recognize that for the breadth and depth of his scientific understanding, for the originality and influence of his scientific work, for the help and support he gave to men like Godfrey and Bartram, Colden and Franklin, he deserves to be known as one of the founding fathers of the scientific tradition in America.

Selection of the Correspondence of Linnaeus and Other Naturalists, ed. James E. Smith (London, 1821), I: 19.

[46] *Historie der Versuche welche vom Jahre 1691 an, bis auf das Jahr 1752 über das Geschlechte der Pflanzen angestellt worden sind* in *Opuscula botanici argumenti*, ed. J. C. Mikan (Prague, 1797), pp. 188–91.

[47] See, for example, Johann Hedwig, *Theoria generationis et fructificationis plantarum* (Leipzig, 1798), p. 44; Auguste P. de Candolle, *Regni vegetabilis systema naturale* (Paris, 1818), I: 71; Ludolf Christian Treviranus, *Die Lehre vom Geschlechte der Pflanzen in Bezug auf die neuesten Angriffe erwogen* (Bremen, 1822), pp. 100, 106.

[48] Oxford, 1906, p. 392. Sachs cited Pritzel's *Thesaurus literarum botanicarum*, in which the *Experimenta* were erroneously said to have been published at The Hague.

President Thomas Clap of Yale College: Another "Founding Father" of American Science†

*By Leonard Tucker**

UNTIL recent years, the customary practice of scholars seeking to recount the scientific accomplishments of colonial America was to summarize the near-legendary feats of Benjamin Franklin.[1] Such a convention was in keeping with the established belief that science was not an important element in the cultural life of colonial America. The corollary of this view was, of course, that Franklin was a cultural freak —"advanced for his time"— because of his strong predilection for science. Both notions have been dispelled through the findings of a small, but growing, legion of historians who have tracked through the uncharted area of early American science. Each year brings additional studies which further illuminate the darkened field of inquiry.

This recent emphasis has already had some important effects. First, a new dimension has been added to our conception of American civilization during its formative period. Science, it now appears, was an integral part of the cultural pattern of colonial America. The intellectual horizons of these early Americans were far broader than had been imagined. A second main consequence has been the amassing of a large corpus of knowledge about early American science. Much has been learned of its character, of the scientific ideas then in vogue, and of the types of instruments used in experiments.

Still another result has been the realization that colonial colleges were centers of scientific activity. The findings of such scholars as Samuel Eliot Morison, I. Bernard Cohen, Lao Simons, Theodore Hornberger, Louis McKeehan, and, most recently, Brooke Hindle have revised the traditional view of these institutions as mere "ministry factories" in which the sciences barely existed.[2]

† I am indebted to the American Philosophical Society for a grant by which I was able to conduct research on President Clap in Great Britain in the summer of 1959.

* Director, Historical and Philosophical Society of Ohio, Cincinnati.

1 Whitfield J. Bell, Jr., anatomizes the reasons for the past neglect of early American science and lists many of the recent books and articles devoted to the subject in *Early American Science: Needs and Opportunities For Study* (Williamsburg, Virginia, 1955), pp. 3-36.

2 Morison, *Harvard College in the Seven-*

Along with the awareness that scientific subjects occupied an exalted position in the curricula of the colleges has come the discovery — perhaps "exhuming" might be a more fitting term — of a number of academic scientists with impressive intellectual abilities. After years of anonymity, David Rittenhouse, William Small, Thomas Robie, Ezra Stiles, Isaac Greenwood, and, most notably, John Winthrop IV have been accorded recognition as important contributors to American cultural growth.[3] At last they have acquired the status of "founding fathers" of American science, an honor formerly held by a meager handful of non-academic figures, chief of whom were Franklin and Jefferson.

Thomas Stephen Clap (1703-1767), Congregational clergyman and for twenty-six years (1740-1766) the president of Yale College, is another academician worthy of inclusion in the Pantheon of early American science.[4] Although he was one of the most influential intellectual leaders of eighteenth-century America, Clap has been consigned to historical limbo, even by the institution he served.[5] If posterity remembers him at all, it is as a bullheaded,

---

teenth Century (2 volumes, Cambridge, Mass., 1936), I, Chapters X, XI; Morison, "The Harvard School of Astronomy in the Seventeenth Century," *The New England Quarterly*, 1934, *7:* 3-24; Cohen, *Some Early Tools of American Science: An Account of the Early Scientific Instruments and Mineralogical and Biological Collections in Harvard University* (Cambridge, Mass., 1950); Simons, "The Adoption of the Method of Fluxions in American Schools," *Scripta Mathematica*, 1936, *4:* 207-219; Simons, "Introduction of Algebra into American Schools in the Eighteenth Century," Bureau of Education, *Bulletin*, no. 18 (1924); Hornberger, *Scientific Thought in American Colleges, 1638–1800* (Austin, Texas, 1945); McKeehan, *Yale Science, the First Hundred Years, 1701–1801* (New York, 1947); Hindle, *The Pursuit of Science in Revolutionary America 1735–1789* (Chapel Hill, North Carolina, 1956), Chapter V. This is not to be taken as a complete list of those who have written on scientific developments in the colonial colleges. Important studies have been made by others, among them Frederick Brasch, Margaret Denny, Frederick Kilgour, Henry Fuller, and Galen Ewing.

[3] See, for example, Maurice Babb, "David Rittenhouse," *Pennsylvania Magazine of History*, 1932, *56:* 193-224; Thomas Cope, "David Rittenhouse — Physicist," Franklin Institute, *Journal*, 1933, *215:* 287-297; W. Carl Rufus, "David Rittenhouse as a Mathematical Disciple of Newton," *Scripta Mathematica*, 1941, *7:* 228-231. Herbert Ganter, "William Small, Jefferson's Beloved Teacher," *William and Mary Quarterly*, third series, 1947, *4:* 505-511; Small spent only six years in the colonies but nonetheless made important contributions to American science. Frederick Kilgour, "Thomas Robie (1689-1729) Colonial Scientist and Physician," *Isis*, 1939, *30:* 473-490. Professor Edmund Morgan of Yale University is preparing for publication a biography of Stiles. Lao Simons, "Isaac Greenwood, First Hollis Professor," *Scripta Mathematica*, 1934, *2:* 117-124. Frederick Brasch, "Newton's First Critical Disciple in the American Colonies—John Winthrop," *Sir Isaac Newton, 1727-1927: A Bicentenary Evaluation of His Work* (Baltimore, 1928), pp. 301-338; Frederick Brasch, "John Winthrop (1714-1779), America's First Astronomer, and the Science of His Period," Astronomical Society of the Pacific, *Publications*, 1916, *28:* 153-170; Clifford K. Shipton, *Biographical Sketches of Harvard Graduates* (Volumes IV-IX, Cambridge, Massachusetts, 1933-1956), Shipton's works are a rich source for the history of science at Harvard. See, in particular, his vignettes of Robie, Greenwood, and Winthrop IV.

[4] Clap became Rector of Yale in 1739, and under the new Charter of 1745 (drawn up by Clap) assumed the title "President." The Charter is printed in Thomas Clap, *The Annals or History of Yale College. . . .* (New Haven, 1766), pp. 45-52. On Clap's life and professional career, see Shipton, *Biographical Sketches of Harvard Graduates*, VII, 27-50; Franklin B. Dexter, "Thomas Clap and His Writings," New Haven Colony Historical Society, *Papers*, V (New Haven, 1894), pp. 247-274.

[5] Yale, which has long maintained a beady eye on its past and been accused by some of carrying tradition "almost to absurdity" (as Professor George Pierson of Yale puts it), has not even accorded Clap the token recognition of naming a passageway or portal in his honor. Such recognition has been conferred upon virtually every important colonial associated with the college. One source affirms that Clap's

contentious Calvinist who almost destroyed Yale when he attempted to convert it into a Congregational seminary. Clap's administration is often regarded as an ugly footnote in the history of Yale, for it was unquestionably the most turbulent era of the school's entire history. Controversy after controversy rocked the college until finally, in 1766, Clap was driven into resignation by a rebellious student body that almost reduced the college buildings to a heap of rubble. Although many factors were involved in the deposal of Clap, one of the most important was his unswerving desire to impose old-style religious values upon a society no longer responsive to the orthodox creed. In matters of faith, Clap was a staunch, uncompromising conservative, of the mettle of the Mathers. He was, in a word, out of season.

Yet, if Clap exhibited a rigid conservatism in religious and philosophical thought, he was a modernist in scientific orientation, an avid supporter of the "new philosophy," a foremost expositor of Newtonian theory. With the substitution of the quadrant for the Bible in his hand, a transformation of his personality seemingly developed. At one moment he was a spokesman for medievalism, at the next a herald of the new world. Such intellectual dualism was not uncommon among the intelligentsia of New Zion. The New England historical landscape is dotted with Puritans who made a similar accommodation between old religion and new science. As Perry Miller has conclusively shown, rationalism was a principal foundation stone in the intellectual structure of Puritanism. Cotton Mather spoke for all Puritans of scientific bent when he wrote: "Philosophy [Science] is no Enemy, but a mighty and wonderous Incentive to Religion."[6]

In Clap's case, the accommodation of religion and science took place at Harvard College. Here he was carefully fed Newton in a scriptural spoon. From all available evidence, his mentor was Tutor Thomas Robie, a devout Puritan and one of colonial Harvard's leading scientific scholars.[7] Robie was well aware that science was a two-way road, one direction leading to Christian truth, the other to skepticism and, possibly, atheism. He saw to it that his students followed the Puritan route; that they accompanied their scientific studies, in the grandiloquent phraseology of Cotton Mather, "with continual Contemplations and agreeable Acknowledgments of the Infinite God."[8] Science was not conceived, and was not taught, as an independent course of study in the Harvard system of education but remained tightly bound with moral and theological precepts. It was regarded more as a method than a philosophy, as a technique of investigation and not an account of being.

Thus natural causation and divine determination became mutually compatible. Science stood as a bulwark to religion, buttressing the "truths" of

name was once recommended for a residential college, but the Yale officials ultimately decided upon Ezra Stiles. In this connection, there may be more truth than humor in the jest of a Yale authority who informed me that naming a building after Clap is "semantically impossible."

6 Kenneth B. Murdock, ed., *Selections From Cotton Mather* (New York, 1926), p. 286. Mather's statement is from *Christian Philosopher.*

7 The regular tutor for Clap's class was Henry Flynt, but Robie, according to Samuel Eliot Morison, "specialized in Science." *Three Centuries of Harvard,* p. 58.

8 *Manductio ad Ministerium* (Boston, 1726), p. 47.

Christianity. If Newton's astronomy and physics enlarged man's conception of the universe, such increment cast greater glory, rather than doubt, upon the Creator. The marvels of science merely reaffirmed the majesty of Deity. Observing the universe through a telescope reveals more than planets, stars, nebulae, and the like; the spectacle proclaims the glory of God. In this spirit did Robie instruct his students. Robie also made the traditional Puritan application of scientific knowledge. Science provided man with a matchless opportunity to gain an insight into the wondrous design of the Creation. Through the knowledge derived, man could better determine his position in God's intended scheme and thereby live in closer accord with the Divine Plan.

The Harvard experience was not forgotten by Clap. In later years, in his more reflective religious writings, he reaffirmed the Puritan tenet that divine revelation was a greater source of truth than unaided reason. God-consciousness remained at the sub-stratum of his scientific thought.

One point, however, is worthy of attention. Clap's scientific outlook, while of Puritan flavor, was not of seventeenth-century vintage; the Yale president moved in quite another world of thought. The contents of his extant scientific writings reflect an outlook far more rationalistic and secular than that of his forefathers.[9] The distinction lay in degree. Clap's writings are conspicuously empty of ecstatic statements on the marvels wrought by God. The overwhelming emphasis is on "how" not "why"; on "cold, irreducible facts" not theological implications; on natural not supernatural explanations. Nor did Clap discuss scientific matters in a religious idiom. He made hypotheses and sought to confirm them through empirical observation and inductive reasoning. He used the methodology, and spoke the language, of the new age.

From the scientific standpoint, Clap was indeed a solid citizen of the world of the Enlightenment. He worshipped at the altar of Isaac Newton, the scientific master-spirit of the age, to whom he extended this tribute: "there are many important Truths in natural Philosophy and Mathematics, which, when they come to be fairly proposed, were never doubted of; such as the general Laws of Attraction, the Weight of the Atmosphere, Rules of Fluxions, etc. and yet it is probable that these Things never came into the Mind of any Mortal, till they were suggested by the great Genius of Sir Isaac Newton."[10] The Newtonian framework provided the foundation for Clap's conceptual scientific scheme, and the *Principia* became his book of scientific revelation. Its theories and rarified principles were as absolute and incontrovertible to him as Puritan religious tenets. "I have known him to elucidate so many of the abstrusest theorems and ratiocinia of Newton," Ezra Stiles wrote, "that, I doubt

[9] The inevitable exception must be acknowledged. The Yale Library has a manuscript sermon by Clap pertaining to the propriety of using natural, rather than artificial, light which suggests an anti-scientific attitude. Clap initially developed the point that the use of artificial light is dangerous to Christian morals. When discussing the advantages of natural light, however, he assumed a more "scientific" attitude, pointing out, for example, that students studying by artificial light who had a tendency to move their eyes from light to darkness would suffer discomfort because of the "instant contractions and dilations." The internal evidence suggests that the sermon was one of Clap's earlier writings.

[10] *An Essay on the Nature and Foundation of Moral Virtue and Obligation* (New Haven, 1765), p. 46.

not, the whole *Principia* of the illustrous philosopher was comprehended by him; a comprehension which, it is presumed, very few mathematicians of the present age have attained."[11]

Clap's traits as a scientist also clearly reveal his rapport with the Enlightenment. Initially, he was imbued with a robust scientific optimism. He was confident, for example, that human reason was capable of penetrating the innermost secrets of nature. (In his religious writings he never placed a transcendent value upon reason.) In at least one instance his optimism carried him to lofty, and, for a Puritan, incredible heights of speculation. He wrote in his *Essay on Moral Virtue:* "It is possible that a Man, by contemplating his own Ideas, the visible Works of Creation, and the Course of Causes and Effects, might by Degrees, find out the Being and Perfection of God."[12]

Complementing this optimism was a tolerant and liberal attitude noticeably absent in his religious personality. While locked in bitter controversy with Jared Eliot and Benjamin Gale, two of his most implacable foes, over issues fundamentally religious in character, he freely contributed his scientific services to assist these men in simplifying the design of an English "wheat plow." In its current design, the plow was too costly for the farmers of Connecticut. At the request of Eliot, Clap applied his "mathematical Learning and mechanical Genius" to the problem of reducing its number of intricate parts while preserving its utility. He successfully completed the task, Eliot writing later that Clap's efforts had resulted in the development of a plow that "can be made with a Fourth Part of what Mr. Tull's will cost."[13] Clap also amiably served with Eliot as a two-man committee representing the Society for the Encouragement of Arts, Manufactures, and Commerce of London; their job was to promote the raising of silk in Connecticut.[14] Clap's spirit of tolerance was displayed by the harmonious scientific relationships he maintained with other men whose religious beliefs differed radically from those of Congregationalism. He carried on a warm correspondence with Peter Collinson, a Quaker; with William Whiston, an Arian, whose heretical utterances had resulted in his being sacked at Cambridge University; with Benjamin Franklin, a professed Deist during most of his adult life; with Cadwallader Colden, an avowed Materialist in religious and philosophical conviction.

Clap's ambivalence is also revealed by the scientific ideals to which he adhered. He implicitly believed in, and practiced, the principles of free investigation of natural phenomena, the unshackled commerce of scientific information, and mutual criticism of theories. Further, his scientific point of view was that of a relativist, not an absolutist. He accepted the fact that scientific "truth" of the present would become historical lumber in the future. Fully aware that his own cherished theory of meteors was but a waypoint on the long road of astronomical science, he prefaced his conclusions in his essay on meteors with these words: "Let us, then, for the present, until we have fur-

[11] Stiles to Naphtali Daggett, 28 July 1767, MS, Yale University Library.

[12] P. 46.

[13] *Essays Upon Field Husbandry in New England and Other Papers, 1748-1762,* Harry J. Carman and Rexford G. Tugwell, eds. (New York, 1934), pp. 116-117.

[14] *Ibid.,* pp. 128-129; *New London Summary,* 12 September 1760; *Pennsylvania Gazette,* 10 August 1758; Thomas Clap to Royal Society of Arts, 2 June 1760 (MS, Royal Society of Arts Library, London).

ther light by more accurate observations. . . ."[15] On the doctrines of original sin or the divinity of Christ there was no room for conjecture. These propositions were eternal truths, confirmed for all time.

In his scientific personality, President Clap showed no traces of Philistinism. His relationship with Cadwallader Colden best illustrates the point. In July of 1753 Clap received some of Colden's scientific treatises,[16] one of which was, assuredly, *The Principles of Action in Matter, the Gravitation of Bodies, and the Motion of the Planets, Explained from those Principles* (London, 1751). In this work Colden sought to extend the Newtonian synthesis by ascertaining the cause of gravitation, no small feat for a provincial scientist; as Professor Hindle has commented, "no more audacious claim to intellectual eminence was ever made in colonial America. . . ."[17] Working in the light of Newtonian celestial mechanics, Colden posited a thesis that pointed to ether as the basic element responsible for gravitational stress. Newton himself had projected this thesis but had not bothered to develop it. Colden gave it a full-dress treatment, but in terminology that did not exactly do honor to clarity.

After working his way through Colden's scientific-metaphysical maze, Clap wrote a letter to the New Yorker in which he discussed his central points. With Colden's philosophical belief that "we have no Knowledge of Substance or any Being upon us," he heartily concurred, a surprising fact in view of the proposition's materialistic overtones. He was skeptical, however, of the validity of the "Idea of [ascribing] a Power or Principle of self Motion, or Agency to a *non* intelligent." Reason, Clap countered, demonstrates that the universe is governed by an intelligent force, namely God. Clap did not state it as bluntly as that, but his meaning was patently clear: "We plainly perceive the *Fact* that Bodies gravitate tend or go towards the Earth, but the efficient *Cause* of this Phenomenon, we do not perceive by our Senses and can only conjecture by our Reason and I can't conceive of any Cause capable of producing any such Effect, but an *INTELLIGENT* one." Be that as it may, he intended to reread Colden's "ingenious Theory." He was certain that his "Conception of those Things, will be much enlarged."[18]

The significance of Clap's sentiments lies in their temperate and critical assessment of a scientific theory that had obvious heterodox implications. Colden, in short, eliminated God from the scheme of things, ascribing inherent activity to matter. As analyzed by I. Woodbridge Riley, his basic theory, combining "the Newtonian mechanics with the ancient hylozoistic doctrine of a cosmic substance as itself intelligence, ultimately reached a kind of dynamic panpsychism, substance being conceived as a self-acting and universally diffused principle, whose essence is power and force."[19] Whatever the implications of Colden's theories, the important point is that Clap read these writings

[15] *Conjectures Upon the Nature and Motion of Meteors Which Are Above the Atmosphere* (Norwich, Conn., 1781), p. 12.

[16] It is interesting to note that Colden's writings had been forwarded to Clap by Samuel Johnson, president of King's College, with whom Clap was feuding violently at the time over religious matters.

[17] "Cadwallader Colden's Extension of the Newtonian Principles," *William and Mary Quarterly,* third series, 1956, *13:* 459.

[18] Clap to Colden, July, 1753, MS, Historical Society of Pennsylvania.

[19] *American Philosophy: The Early Schools* (New York, 1907), p. 21, recently reprinted by offset and reissued by Russell and Russell, Inc., New York.

to acquire understanding, not to confute, as was the case when he studied religious tracts of a non-Congregational substance. His mind was not hedged by defensive barriers. Only in science was he receptive to the novel and the strange.

In the prevailing fashion of the literati of the Enlightenment, Clap dabbled in diverse types of scientific activity. All the world of science was his domain; as Whitfield Bell Jr. writes of eighteenth-century science, it "was not yet compartmentalized, each tight little box jealously guarded by guildsmen who talked to themselves and one another in a private jargon."[20] Clap's scientific catholicity is spectacular: he maintained temperature charts,[21] devised agricultural tools, engaged in silk-raising, speculated on the medicinal value of mineral springs,[22] and investigated sun spots, transits of Mercury, eclipses, comets, and meteors.

His Baconian breadth of interest was matched by a passionate desire to be well informed on recent developments, especially in astronomy. He kept abreast, primarily through correspondence, with some of the more distinguished scientific virtuosi, both English and colonial, including Franklin, Colden, Peter Collinson, Cromwell Mortimer, John Pringle, and William Whiston.[23] Moreover, as revealed in his writings, he read extensively in the standard scientific works of the day; for astronomical tracts, in particular, he had, in the idiom of Jefferson, a "Canine Appetite." The *Gentleman's Magazine,* and the *Philosophical Transactions* of the Royal Society, kept him informed of developments in England and elsewhere.[24] His personal library contained a generous portion of scientific works,[25] and, of course, he had access to the rich collections in the college library.

[20] *Early American Science,* p. 9.

[21] Franklin B. Dexter, ed., *Extracts From the Itineraries and Other Miscellanies of Ezra Stiles* (New Haven, Conn., 1916), p. 221, hereafter cited as Dexter, ed., *Stiles' Itineraries.*

[22] In 1765, a Connecticut resident afflicted with a severe skin disorder was miraculously cured after bathing in the mineral springs of Stafford Springs (Connecticut). Overnight Stafford became the "Mecca of New England's hypochondriacs." Clap bathed in the springs in 1766 and came away convinced of their therapeutic value. In a letter to the editor of the *Connecticut Courant* (18 August 1766), he called upon physicians and learned men from the Stafford area to meet in conference and formulate a program whereby the springs would be exploited in the most effective manner. As part of an eight-point plan, he proposed that experimentation be conducted and the results announced to the general public. Additionally, he urged that a skilled physician be maintained at the springs to advise the "unskilled Multitudes." Clap became concerned over the improper use of the springs and the fraudulent practices that had developed. (The *Boston Gazette,* 4 August 1766, noted this case. A group of invalids in Boston hired a carter to haul water to them from Stafford. En route to Boston, the carter began to sell the water. When the barrels ran dry, he refilled from the nearest brook and later boasted of a turnover of some 160 gallons. His unsuspecting clients in Boston drank the brook water and, in their collective opinion, benefited immeasurably from it.) Clap's letter was reprinted in the *Boston Post-Boy,* 6 September 1766. On the mineral springs of Stafford, see Carl Bridenbaugh's delightful article, "Baths and Watering Places of Colonial America," *The William and Mary Quarterly,* third series, 1946, *3:* 152-158.

[23] Clap constantly sought to widen his relationships with English scientists. In a letter to William Whiston, 1 July 1752 (MS, Yale University Library), he requested the names of some "ingenious Astronomical Gentlemen in London" with whom he might correspond. In the summer of 1959, I made an intensive search of Scottish and English depositories but was unable to locate any of Clap's letters.

[24] The college library had a nearly complete set of these two collections.

[25] Clap's book list is contained in Probate Records of Thomas Clap, 14 February 1768, Connecticut State Library. The segment of his library on science was top-heavy in books on astronomy.

A recounting of Clap's scientific activities must begin with a synopsis of his accomplishments as a college administrator and classroom teacher. Before Clap's appointment as rector, science had been largely a peripheral study at Yale.[26] In the early history of the college, particularly in the period 1701-1716, little heed was paid to mathematics and natural philosophy. Although these studies were included in the curriculum, the conservative-minded trustees who regulated the curriculum and prescribed its subject matter adhered closely to the medieval convention of emphasizing moral philosophy, divinity, and the "tongues" (principally Latin). The curriculum became mired in the bogs of scholasticism. The little mathematical and scientific instruction that was provided remained medieval in theory and content.[27] Aristotle supplied the principles of physics, and Ptolemy provided the cosmology. In scientific outlook Yale was on an intellectual plane with Dante. Samuel Johnson, a tutor in the years 1716-19, wrote impatiently that the curriculum of these early years was "nothing but the scholastic cobwebs of a few little English and Dutch systems that would hardly now be taken up in the street."[28]

Johnson's appointment as tutor in 1716 marks the first important departure in the teaching of mathematics and science at Yale. It was he who shook Yale out of its intellectual torpor. During his senior year (1714), the college received a collection of books gathered together by Jeremiah Dummer, New England's colonial agent in London, who was an avid supporter of the young school. Among the collection were some of the chief writings of the English scientific galaxy: Newton's *Principia* and *Optics,* and the works of Edmund Halley, William Whiston, Richard Bentley, Robert Boyle, and Francis Bacon.[29] These works furnished Johnson's famished intellect with a "feast of fat things." His tutors, orthodox to the marrow, cautioned him against the toxic effects of such rich reading. They argued against these new scientific notions because, as Johnson explained, "the new philosophy it was said would soon bring in a new divinity and corrupt the pure religion of the country." Johnson disregarded their admonitions. When he became a tutor, he swept out much of the accumulated scholastic rubbish and introduced the empiricism of Locke, the inductive method of Bacon, and the new cosmology of Kepler, Copernicus, and Newton. He also introduced the vitally necessary adjunctive mathematical studies, although he was obliged to undergo an intensive "do it yourself" course of instruction to acquire a teachable knowledge of "Euclid, Algebra and Conic Sections"; such instruction had not been provided during his own undergradu-

[26] Scientific developments in the early history of Yale are discussed in McKeehan, *Yale Science,* and Edwin Oviatt, *The Beginnings of Yale, 1701-1726* (New Haven, 1916).

[27] One example is in order. During Rector Pierson's administration (1701-1707) the students were taught that sleep was caused by "steames of food, and blood ascending into the Brain, by whose coldness they are said to be condens'd into moisture, which obstructs the passage of the Spirits that they can't freely permeate to the Organs of Senses." Oviatt, *Beginnings of Yale,* pp. 273-276.

[28] Herbert and Carol Schneider, eds., *Samuel Johnson—President of King's College—His Career and Writings* (4 volumes, New York, 1929), I, 5-7, hereafter cited as Schneider, ed., *Writings of Johnson.*

[29] For an analysis of this profoundly significant collection, see Thomas G. Wright, *Literary Culture in Early New England, 1620-1730* (New Haven, 1920), pp. 184-187; see also Louis Shores, *Origins of the American College Library, 1638-1800* (New York, 1935), pp. 75-80, 127-135, 218, 223.

ate career.[30] During his three-year tutorial stint, Johnson was eminently successful in establishing an up-to-date "bookish" scientific tradition.[31] Lacking "philosophical apparatus," he was unable to make progress in experimental science.

After Johnson left Yale in 1719, his teachings were carried forward by Rector Timothy Cutler and Tutor Daniel Brown (sometimes Browne), both of whom had also made a surreptitious examination of the books in the Dummer collection, especially the sermons of Anglican divines. But with the celebrated Cutler defection of 1722, all of the Newtonian-influenced members of the teaching staff were eliminated from the college.[32] Science and mathematics continued to be stressed during the tutorship of Jonathan Edwards (1724-1728), who was also a disciple of Newton, but with less secular emphasis. During the rectorship of Elisha Williams (1726-1739), the most notable scientific progress was made in the accumulation of laboratory apparatus; until 1734 the apparatus consisted of but two globes.[33] Holding no deep-seated interest in mathematics and science, Williams gave them little stress. As one authority has written, his influence was "trivial."[34]

Clap's era was the golden age of mathematics and science in Yale's colonial history. The president was individually responsible for a remarkably advanced mathematical and scientific program. It was during Clap's administration that students came to "woo the skeleton of science," as John Trumbull wrote in his slashing attack upon the Yale curriculum.[35]

Clap made significant curricular adjustments shortly after he assumed office. Mathematical subjects leaped into prominence, since Clap, like Samuel Johnson, was of the conviction that youth should begin their collegiate education with the "Languages and the Mathematics (which are themselves indeed a kind of Language) for these are both of them a necessary Furniture in order to the attainment of any considerable Perfection in the other parts of the Learning."[36] Under Rector Williams, the students had received their first mathematical instruction during the senior year.[37] After 1741, however, such

[30] Schneider, ed., *Writings of Johnson,* I, 7-9.

[31] McKeehan, *Yale Science,* p. 9.

[32] Rector Cutler, Tutor Brown and five prominent Congregational ministers, one of whom was Samuel Johnson, abjured Congregationalism and declared their allegiance to the Church of England in 1722. The events may be followed in the documents contained in Francis L. Hawks and William S. Perry, eds., *Documentary History of the Protestant Episcopal Church in the United States of America Containing Numerous Hitherto Unpublished Documents Concerning the Church in Connecticut* (New York, 1863), I, 62-80.

[33] Williams purchased surveying instruments, a reflecting telescope, a microscope, a barometer, and some "mathematical instruments." The globes had been sent from England by Dummer in 1716. See Henry M. Fuller, "The Philosophical Apparatus of Yale College," *Papers in Honor of Andrew Keogh* (New Haven, 1938), pp. 164-165.

[34] McKeehan, *Yale Science,* p. 14. The commencement theses during Williams' rectorship contain a large number of propositions on mathematics and physics. Competent authorities have affirmed, however, that the quality of instruction could not have been of a high order if these theses are to be taken as an indication of curricular content.

[35] "The Progress of Dulness," in *The Poetical Works of John Trumbull* (Hartford, 1820), II, 17. Trumbull was obversely arguing for an emphasis upon *belles lettres.*

[36] Johnson, *An Introduction to the Study of Philosophy* (New London, 1743), advertisement.

[37] Florian Cajori, *The Teaching and History of Mathematics in the United States* (Washington, D.C., 1890), p. 31.

instruction was offered in the freshman year. Clap's course of study in 1743 listed the following mathematical subjects: "arithmetic and algebra" in the first year, geometry in the second year, and advanced mathematics (algebraic conics and fluxions) in the junior year.[38]

The constancy of stress on mathematics is shown by a second curriculum listed by Clap in 1766. "Mathematics" was offered to the freshmen, trigonometry and algebra to the sophomores, and "most branches of the Mathematiks" to the juniors. A parenthetical assertion by Clap offers additional evidence that mathematics was a prominent study: "Many of them well understand Surveying, Navigation and the Calculation of the Eclipses; and some of them are considerable Proficients in Conic Sections and Fluxions."[39] The commencement sheets are a final proof.[40] Propositions relating to all phases of the discipline, from simple algebra to complex fluxions, are listed.[41] In a great many sheets, the mathematical propositions represent the greatest numerical proportion. This fact cannot be overstressed inasmuch as the sheets serve as a barometer for ascertaining the pressures of the curricular "climate of opinion." While some contemporary clerical leaders—John Wesley, for example—inveighed against instruction in mathematics, claiming it set youth straight on the path to atheism, Clap provided his students with an abundance of such instruction, never for a moment entertaining the fear that he was directing youth into irreligion.

Clap also promoted natural philosophy, and by the 1750's it was the aristocrat of the secular portion of the curriculum. While a number of scientific areas were touched upon, particular focus was placed upon physics and astronomy, Clap's personal favorites. The commencement sheets amply reveal the emphasis upon these subjects. The commencement sheet of 1751, for example, listed one hundred and twenty-five propositions, of which thirty-eight were placed under the rubric *Theses Physicae*. Of these, at least one-third more properly dealt with astro-physics than with physics *per se*. The commencement sheet of 1765 contains forty-seven propositions on physics, more than twice the number of any other category.

The most conspicuous features of the astronomical propositions were their variety in range of investigation and their interdependence with mathematics and, to a lesser extent, theology. These few examples lend color to the point. In 1752, a proposition held: "Annis 25,919, Poli Aequatoris, circà Polos Eclip-tae, revolvunt." In 1753: "Cometae in Ellipsibus longè excentricis, maximè appropinquantibus Parabolis, circa Solem revolvunt." In 1754: "Cometae Usui totius hujusce Systematis inserviunt." In 1755: "Si Sol circa Terram quotidiè revolvatur, Velocitate 353,571 Milliarium, uno moveatur minuto." In 1764: "Planetarum a sole Distantia, et earum Diametri sunt ut sequens exhibit Tabella.

[38] *Catalogue of Yale Library* (New London, 1743), advertisement.

[39] *Annals of Yale,* p. 81.

[40] The Sterling Memorial Library (Yale University) has a nearly complete collection of commencement sheets for Clap's administration.

[41] Fluxions (or integral calculus) first appeared in the commencement sheet of 1758. In the judgment of Lao Simons, the problems became increasingly difficult with the passing of years. "Adoption of Fluxions in American Schools," *Scripta Mathematica,* 1936, *4:* 207-210.

	Distantia	Diameter
Mercurii	37,090,000	3,596
Veneris	69,290,000	9,285
Terrae	95,820,000	7,897
Martis	146,000,000	5,275
Jovis	495,000,000	95,319
Saturnii	913,800,000	79,828"[42]

Student astronomical almanacs,[43] the comments of such qualified contemporaries as Samuel Johnson and Bishop George Berkeley,[44] and student manuscripts and commonplace books[45] further indicate that Clap charged the Yale youth, in the rhetoric of Cotton Mather, "to soar upwards, to the Attainments of ASTRONOMY."[46]

The success of a program of experimental science was contingent upon the amount and quality of available apparatus. This was especially true in the areas of physics and astronomy, which, for proper study, required such basic instruments as globes, thermometers, quadrants, and telescopes. When Clap came to Yale there was but the closetful of equipment accumulated by Rector Williams. He set himself to the task of augmenting this meager collection. His standard technique was letters of appeal directed to wealthy or influential colonials and Britons known to have an interest in science.[47]

At times the pleas were subtle (or at least were intended to be). On 31

[42] Translated, the propositions read:

1752—In 25,919 years the poles of the equator revolve around the poles of the ecliptic.

1753—Comets revolve around the sun in eccentric ellipses at a great distance and in parabolas which approach the sun as closely as possible.

1754—Comets serve the use of the whole system.

1755—If the sun revolves around the earth daily, it would be moved with a velocity of 353,571 miles a minute.

1764—The distance of the planets from the sun and the diameters of them are as the following tables show.

[43] Joseph Huntington, *College Almanack, 1761. An Astronomical Diary* (New Haven, 1761); Huntington, *College Almanack, 1762. An Astronomical Diary* (New Haven, 1762).

[44] Berkeley to Clap, July 17, 1750, *Yale University Library Gazette*, VIII (July, 1933), p. 28. Schneider, ed., *Writings of Johnson*, I, 102.

[45] Mannaseh Cutler, *Commonplace Book Kept at Yale From June 10, 1762*, typed copy in Library of Congress; Cutler, *Book of Astronomical Recreations, Performed at Yale College, New Haven, A.D. 1763*, MS, Essex Institute; Trumbull, *Poetical Works*, I, 11; Eleazar May, *Commonplace Book*, MS, Yale University Library. In a letter to Cromwell Mortimer (1 April 1744, MS, Royal Society, London), Clap stated that he had set one of his pupils to the task of tracing the trajectory of a comet he had been observing, which suggests his collaboration with students in astronomical research, assuredly in an instructional capacity. The manuscripts cited above provide us with the additional information that Clap adopted such up-to-date texts as Willem Jacob Van s' Gravesande's *Mathematical Elements of Natural Philosophy Confirmed by Experiments; or, an Introduction to Sir Isaac Newton's Philosophy* (translated by J. T. Desaguliers [London, 1720-1721]); William Whiston, *Astronomical Principles of Religion* (London, 1717); William Derham, *Astro-Theology: Or a Demonstration of the Being and Attributes of God, From a Survey of the Heavens* (London, 1715). These works presented Newtonian theory within the traditional Puritan framework.

[46] Mather, *Manductio ad Ministerium*, p. 53.

[47] Some colonial colleges eschewed correspondence and sent solicitors to England. See Edward P. Cheyney, *History of the University of Pennsylvania, 1740-1940* (Philadelphia, 1940), pp. 61-67; William L. Sachse, *The Colonial American in Britain* (Madison, Wisconsin, 1956), p. 112; Beverly McAnear, "The Raising of Funds by the Colonial Colleges," *Mississippi Valley Historical Review*, 1951-52, *28:* 591-612.

May 1743, Clap began a correspondence with Cromwell Mortimer, the secretary of the Royal Society, who was in a strategic position to solicit aid for the Connecticut college. Clap opened his letter with a brief account of a transit of Mercury he had witnessed through the college-owned telescope, an instrument thirty inches in length and four inches in diameter. Next he related his cryptic impressions of some sun spots he had seen on the 28th and 29th of October. He concluded this portion of his letter by mentioning that he had intended to make some critical observations of an eclipse of the moon and of other perturbations but had been prevented from doing so by the inopportune appearance of clouds. Clap's statements, it was plain to see, were designed to impress more than inform; to whet, not glut, Mortimer's astronomical appetite.

President Clap then moved to the heart of the matter. "As we are yet but an Infant College, of scarce 40 years Standing in a new Count[r]y, and destitute of mathematical Instruments, particularly an Orrery, an astronomical Quadrant and an Air Pump, It is not as yet to be expected that we should make much Proficiency in Astronomy, and some other Parts of the Mathematicks." Continuing in this apologetic vein, he reported the lack of an instrument "by which I can take the exact Distance of a Planet, or Spot from the Center or Limb of Sun," and complained that he could not see "above Half the Body of the Sun at once in our Telescope, so that what I have said about the Distance is only guess'd." As for books, while the college had a few basic works on astronomy, it owned no planetary tables, which were indispensable for astronomical research. In a final shot, Clap informed Cromwell that the "B[o]rometer Tube is broke," and he seriously doubted that any man in the colonies could repair it. "Thus hoping for your Smiles upon our young Nursery of Learning," Clap ended his letter.[48]

The following year, after the dramatic appearance of a comet, he renewed his efforts to win Mortimer's assistance in procuring apparatus. "I suppose the late Remarkable Comet has engaged the Attention and Critical Observations of all the Astronomers in Europe," he reflected. "We in the Infant College are under no manner of Advantages for such a Purpose having no Astronomical Quadrant, nor any kind of Instrument adapted to take the angular Distance of any of the heavenly Bodies. . . ."[49] There could be no mistaking Clap's purpose.

Whether these communications brought "Smiles" from Mortimer is not known, but similar pleas apparently did result in the acquisition of some new equipment. In 1757, for example, a London patron of science presented some astronomical instruments to the college. A few years later Philip Schuyler, the New York land baron and Alexander Hamilton's future father-in-law, contributed an unspecified "electrical instrument," possibly a Leyden jar.[50] Still

[48] MS, Royal Society.

[49] See footnote 45. Clap also wrote that he "had thoughts" to raise a subscription to purchase scientific instruments. He inquired of Mortimer the prices of certain items. That the American colonies were deficient in even the most fundamental of scientific equipment is revealed by Clap's action of sending the college's telescope to England in 1750 for a speculum. *Yale Corporation Records,* I, 94, typed copy in Secretary's Office, Yale University.

[50] Franklin B. Dexter, ed., *The Literary Diary of Ezra Stiles* (3 volumes, New York, 1901), II, 348. The Leyden jar demonstrated how static electricity could be generated by means of friction.

another benefactor was Benjamin Franklin, who maintained a warm scientific relationship with Clap and occasionally advised him on new types of apparatus being developed in England.[51] The peripatetic Philadelphian contributed a copy of his celebrated experiments on electricity and a Leyden jar.[52] These philanthropic acts may have played a part in Yale's decision to award Franklin an honorary M.A. degree in 1753, although the Corporation Records maintain, plausibly enough, that the degree was granted in recognition of Franklin's "ingenious experiments and theory of electrical fire."[53]

Franklin was indirectly responsible for the dissemination of scientific information at Yale. When he came to Yale to accept his degree, Clap approached him on the possibility of establishing a printing office in New Haven. The financial barriers for such an undertaking were far too imposing for the local residents. In subsequent correspondence, the president vigorously pursued the project. Finally convinced that the venture represented a promising investment, Franklin consented to undertake it.[54] The valuable press arrived from England in the fall of 1754, and by Christmas the office was open for business.[55] The press was utilized for publishing the annual commencement sheets and masters' *quaestiones*, both of which contained a liberal number of propositions dealing with science. The local townspeople benefited from Franklin's largesse in the form of the *Connecticut Gazette*, New Haven's first newspaper. In this journal Clap published reports of his astronomical observations.

Clap's various maneuvers to gain scientific apparatus for Yale were successful. In a volume of land records begun in 1747, the president listed the "Mathematical Instruments belonging to college:"[56]

> A Telescope with a Tripod; two Setts of Posts and a Glass to be [lowered?] on to look on the Sun.
> A Pair of Globes, Celestial and Terrestrial with Quadrants of altitudes.
> A Pair of old Globes.
> A Theodolite with a Tripod, plain Table, and Brass Scale and Sights for it needles and Glasses.
> Two measuring Wheels.
> A Gunters chain.
> A short wooden Scale.
> A Pair of Dividers.
> A protractor.

[51] In 1753, for example, Franklin, at the request of the President, furnished information on a new air pump developed in England. He urged Clap to secure the model for the college. John Bigelow, compiler, *The Writings of Benjamin Franklin* (12 volumes, New York and London, 1904), III, 169-170.

[52] In earlier correspondence, Clap informed Franklin that he planned to raise money to provide Yale with a "compleat Apparatus for Natural Philosophy." Franklin promised to contribute the "Electrical Part" if Clap was successful in the plan; Franklin to Clap, 28 November 1751, American Philosophical Society, *Proceedings*, January to December, 1895, *34:* 484. The "friction machine" (Leyden jar) is currently housed in the Franklin Room of the Sterling Library, where the papers of the Philadelphian polymath are being assembled for publication.

[53] I, 101.

[54] Bigelow, compiler, *The Works of Benjamin Franklin*, X, 267.

[55] James Parker to Jared Ingersoll, 14 March 1768, Winnifred Reid, "Beginnings of Printing in New Haven," *Papers in Honor of Andrew Keogh*, pp. 67, 70, 79.

[56] MS, Yale University Library.

A Loadstone set in brass with Steel Arms.
A Microscope with [?] Apparatus.
A Barometer and Thermometer.
An Orrery.
A concave Glass.
A curve Glass.
A multiplying Glass.
A Pair of small neat Ballances, or Scales with all proper Weights.
A Landscape Box.
Two Prisms, with a Stand.
A brass Syrings.
About Ten glass Tubes.[57]

While some of these instruments had been procured by Williams, most were acquired during Clap's administration. The President's solicitations apparently continued to inspire philanthropy. In 1779, Ezra Stiles, who had a passion for listing odd bits of trivia in his diary, noted down the apparatus owned by the college.[58] According to I. Bernard Cohen, this collection, while not as extensive as Harvard's, "was clearly well-chosen and of high quality."[59] The bulk of the items on Stiles's list had been procured during Clap's term of office.

There was one piece of equipment that Clap was not able to acquire through donation, but which was essential for astronomical instruction. This was an orrery, an instrument that illustrated the known major bodies of the solar system in their relative sizes and positions. In 1744, Clap put his inventive abilities to work and constructed one. Although crude in comparison with the mechanical wonder built by David Rittenhouse,[60] it was the first of its kind to be constructed in the American colonies.[61] The machine was seven feet in diameter; its component parts were held in place by pins and wires. Metal plates represented the earth, sun, moon, and the major and secondary planets, and even as recent an arrival as the celebrated Comet of 1682 was shown.[62] Unlike the more expensive and more elaborately constructed European models, whose movements were controlled by one central lever, Clap's orrery necessitated hand movements of all the various heavenly bodies and corresponding arithmetical calculations for precise positioning. Operation of the orrery, therefore, provided the students with intensely practical mathematical and astronomical training. Though "rough and unpolished," it permitted them

[57] For a description, and discussion of the function, of such instruments, consult Cohen, *Early Tools of Science.* Many are pictured in this work.

[58] Dexter, ed., *Literary Diary of Stiles,* II, 348-349.

[59] *Early Tools of Science,* p. 10.

[60] Of Rittenhouse's orrery, Jefferson wrote in his *Notes on Virginia:* "He [Rittenhouse] has not indeed made a world; but he has by imitation approached nearer its Maker than any man who has lived from the creation to this day." Paul L. Ford, ed., *The Writings of Thomas Jefferson* (10 volumes, New York, 1892-1899), III, 169.

[61] Howard C. Rice, Jr., *The Rittenhouse Orrery, Princeton's Eighteenth Century Planetarium* (Princeton, New Jersey, 1954), pp. 12-13. Harrold Gillingham, in his article "First Orreries in America," *Franklin Institute Journal,* 1940, *229:* 81-99, fails to mention Clap's machine, crediting Rittenhouse with the honor of constructing the first orrery in America.

[62] Chauncey Whittelsey, "A Description of an Orrery or Planetarium in the Library of Yale-College in New Haven, lately projected and made by the Rev. Rector Clap, to represent the Motions of all the celestial Bodies," *The American Magazine and Historical Chronicle* (January, 1743/4), I, 202-203.

to witness, with a sweep of the eye, the complex movements of the earth and planets and offered a visual explanation for such natural phenomena as eclipses and the reappearance of comets. That the machine was a great stimulus to astronomical research seems certain.

Clap was too perceptive an administrator to believe that mathematical and scientific instruction would improve once the curriculum was properly stocked and philosophical apparatus secured. He was acutely aware of the need for specialized instruction in such technical areas of knowledge. It seems more than coincidence that a strikingly high percentage of the twenty-eight tutors he appointed exhibited proficiency in mathematics and natural philosophy. Such men as Thomas Darling, John Whiting, Ezra Stiles, Elizur Goodrich, Nehemiah Strong, Richard Woodhull, and Punderson Austin were decidedly scientific-minded.[63]

Perhaps of greater significance, there is evidence suggesting that Clap initiated a sort of "teacher training" program to assure a competent faculty. Yale tutors, it is to be remembered, were recent graduates of the college who served short terms, usually two to three years, and then moved on to other professions; tutoring was conceived as an interim career. Tutorial candidates apparently underwent a special, more intensive, course of preparation while students. Ezra Stiles, one of Clap's more brilliant students and himself a competent Newtonian scientist, wrote that Clap "always spoke to the Person on whom he set his eyes for Tutor and desired him to adapt his Studies preparatory."[64] There can be little doubt that prospective tutors received a concentrated dosage of mathematics and science during their senior year when they studied directly with Clap.

Apart from his duties as a teacher of science, President Clap conducted scientific research. He was an active member of an unorganized, but surprisingly close-knit, group of academic scientists in colonial America. Like such accomplished academicians as John Winthrop IV and William Small, Clap concentrated largely on the scientific trinity of mathematics, physics, and astronomy. As a rule, only the academic group showed proficiency in these subjects. Natural historians could delineate with ease the characteristics of flora and fauna, but, lacking a wide grasp of mathematics, they were unable to do research in the lofty heights of physical astronomy. During the colonial period, it was the academic scientist who conducted the most serious, as well as the most productive, research in physics and astronomy.

Of Clap's abilities as a physicist, there is little direct evidence for a definitive evaluation. From his extant scientific writings, the comments of qualified contemporaries, and the character of the theses appearing in the annual commencement sheets, there is strong indication that he was well versed in the principles of Newtonian physics, especially as they applied to astronomy.

As for his mathematical abilities, the evidence is conclusive: He was of Olympian stature in colonial America. His mathematical competence was

[63] A complete list of the tutors, with their dates of service, is appended to Clap's *Annals of Yale*, pp. 92-93.

[64] Dexter, ed., *Literary Diary of Stiles*, II, 514.

openly acknowledged by contemporaries. The reliable Stiles wrote: "In mathematics and natural philosophy, I have not reason to think that he was equalled by any man in America, except the most learned Professor Winthrop. Many others excelled him in the mechanic application of the lower branches of the mathematics, but he rose to sublime heights, and became conversant in the application of this noble science to those extensive laws of nature, which regulate the most extensive phenomena, and obtain through the stellary universe."[65] At another time, Stiles affirmed that Clap surpassed Elisha Williams "as well as all the Presidents in Harvard College, and in all American colleges in Mathematics, Natural Philosophy and Astronomy."[66] Friend and foe alike shared the belief that the President was, as Samuel Johnson understated it, "much of a mathematician."[67] Chauncey Whittelsey, a Yale tutor who held Clap in something less than esteem, echoed Stiles's judgment when he wrote that "in Mathematical and Philosophical Learning [Clap] was neither surpassed nor equalled by any man on this Continent, except Professor Winthrop."[68] Clap alone of the instructional staff was capable of teaching Newtonian fluxions at Yale, a subject he had introduced there. When he left the college in 1766, such instruction was temporarily suspended for want of a qualified teacher.

One of the greater ends to which mathematical instruction was directed at Yale was astronomy, the queen of the sciences during the eighteenth century and Clap's major specialty. "He delighted to survey the Heavens and travel among the Stars," Naphtali Daggett recalled in his sermon at Clap's funeral, "and calculate their wonderfully regular Motions, devoutly entertained with the surprizing Displays of the power and wisdom of the great Creator appearing therein."[69] The bulk of the President's personal research focused on this subject, and part of his effort was devoted to solving the central mystery of eighteenth-century astronomical science: the size of the solar system. According to contemporary astronomers, the key was to be found in the determination of the solar parallax, that is, determining the mean distance of the earth from the sun. Once this figure was established, the way would be clear to calculate the scalar dimensions of the universe through the application of Kepler's third law. The "frame of the world" would then be fixed, the cosmological synthesis of Newton would be complete. Solar parallax, as one authority on eighteenth-century astronomy has written, "becomes the standard measure of the universe, a kind of celestial meter stick of no less importance than its terrestrial counterpart."[70] And how was the solar parallax to be determined? In the early years of the eighteenth century it was thought that the infrequent transits of Mercury and Venus provided the celestial mechanism for ascertaining the needed value.[71]

Clap was very much alive to the currents of astronomical thought running

[65] See footnote 11.

[66] Dexter, ed., *Literary Diary of Stiles,* II, 336. Stiles was "personally acquainted" with sixteen college presidents. *Ibid.,* II, 335.

[67] Schneider, ed., *Writings of Johnson,* I, 102.

[68] Dexter, ed., *Stiles' Itineraries,* p. 561.

[69] *The Faithful Serving of God* (New Haven, 1767), p. 32.

[70] Harry Woolf, *The Transits of Venus—A Study of Eighteenth-Century Science* (Princeton, New Jersey, 1959), p. 3.

[71] *Ibid.,* pp. 27-29, 71.

between England and continental Europe. When a transit of Mercury took place in October 1742, he was at his post. Through the thirty-inch reflecting telescope belonging to the college, he observed the passage of the planet across the face of the sun. Observation was the full extent of his activity, for he lacked the necessary instruments to make mathematical computations.[72] As the time for another transit of Mercury neared in 1753, Clap made "great Preparations" to view the spectacle. Now he was equipped to make calculations. A quadrant and an astronomical clock lay close at hand. The telescope stood poised.[73] One unpredictable element, however, thwarted his efforts — when he arose before sunrise on May 6, heavy clouds blanketed the Atlantic seaboard, frustrating all the colonial astronomers who had carefully prepared for the event.[74] Only one astronomer from the British colonies in America, an observer in Antigua, got a clear view of the transit. Clap soon learned of his "very accurate observation" and anxiously looked forward to receiving more detailed information.

While it is certain that Clap maintained a keen interest in the transit of Venus in 1761, the top astronomical happening of the century, there is no evidence to indicate his active participation in the enormous cooperative scientific enterprise that accompanied the event. Assuredly, Clap was envious of Winthrop, whose fortune it was to receive the generous financial support of the Massachusetts government in outfitting a scientific expedition to St. Johns, Newfoundland, where the transit could be observed; it could not be seen in New Haven because it occurred before sunrise there.

A concatenation of circumstances, then, from a lack of equipment to the vagaries of weather, thwarted Clap's desire to assist in the resolution of the problem of scalar dimension. His sole contribution to the enterprise was that of educating those around him on the significance of this research. In the cultural wilderness of colonial Connecticut, such a contribution was of no mean proportions. Education of this sort, however ephemeral it may appear by modern standards, was prerequisite for the establishment of a tradition of astronomical research.

A second phase of Clap's work in observational astronomy involved the tracking of comets. He shared the general excitement of astronomers in the last months of 1758 over the impending appearance of Halley's Comet.[75] This was to be their sole opportunity to witness one of the most spectacular and mysterious of all natural phenomena, for the comet came within telescopic sight of the earth but once in a span of 76-77 years. The precise date of its appearance was not known. Halley had predicted just prior to his death in 1742 that it would return from the unknown and reach its perihelion in late

[72] Clap to Cromwell Mortimer, 31 May 1743. In this same period, Clap observed sun spots, an eclipse of the moon, and Saturn's rings.

[73] See Clap to Colden, July, 1753 (footnote 18).

[74] I. Bernard Cohen, "Benjamin Franklin and the Transit of Mercury in 1753," American Philosophical Society, *Proceedings,* 1950, *44:* 222-232.

[75] After calculating the orbit of the Comet of 1682, Edmund Halley (1656-1742) predicted that this particular phenomenon would reappear at intervals of approximately 76-77 years. When his prediction was borne out in 1759, the comet was named for him. For a discussion of Halley's Comet, see George Chambers, *The Story of Comets* (Oxford, 1909), Chapter 9.

May 1759. French astronomers, after correcting errors in Halley's mathematical computations, predicted that its perihelion would occur at least one month earlier.[76] Whether Clap had knowledge of these revisions cannot be determined. The close ties between the members of the European astronomical circle, coupled with the fact that Clap corresponded with English astronomers, would suggest that he had been informed of developments concerning the comet. At any rate, it is certain that Clap took up his station and spent many anxious moments peering at the heavens from the end of 1758 through the early months of 1759. John Winthrop did likewise at Cambridge, as did Ezra Stiles at Newport and Theophilus Grew at Philadelphia.

Initial detection of the comet was made by astronomers in continental Europe during Christmas week, 1758. It was first seen in the American area by sharp-eyed watchers in Charleston, South Carolina, in late March 1759.[77] Two weeks later the New England observers caught sight of it. Winthrop tracked it in his telescope on 3 April, and an account of his observations appeared in the Boston newspapers one month later.[78] In early April, Clap caught sight of the fiery comet, whose unique tail of gas molecules, according to modern astronomers, stretches to the phenomenal length of some ninety million miles. One month later he published his observations in the *Connecticut Gazette,* beginning dramatically, "The long expected comet now appears."[79] Nine days later the *Boston Evening Post* reprinted the account.[80] According to Clap's computations, the comet was approximately forty-seven million miles from earth when it reached perihelion and it came closest to earth on 26 April. Interestingly, Clap's major generalizations, derived through independent research, approximated those of the renowned Winthrop.

Clap's third area of astronomical specialization, the study of meteors, was closely connected with his work on comets. He had been conducting research on meteors since his ministerial days in Windham, Connecticut. In the 1750's he announced to astronomical colleagues in America a theory that certain meteors had the main characteristics of comets, in that they made elliptical orbits and did not fall to earth as meteorites or dissipate into gas. Clap's theory, in brief, embodied these points: "superior Meteors," or terrestrial meteors, were solid bodies about one-half mile in diameter that revolved around the earth, their center of focus; the meteors were twenty to thirty miles from the earth at perigee; through friction with the atmosphere they made a constant rumbling noise, similar to earthquakes, and achieved a brilliant luminosity; explosions that usually accompanied these meteors were caused by spark discharges resulting from overcharging; their rate of speed was 500 miles per minute.[81]

While the fundamental theory of terrestrial meteors may have been original with Clap, its seminal idea was first set in print by an unknown contributor[82] to the *Gentleman's Magazine* in October 1755. After giving an account of a

[76] McKeehan, *Yale Science,* p. 20.

[77] *Boston News Letter,* 3 May 1759.

[78] *Ibid.; Boston Evening Post,* 7 May 1759.

[79] 5 May 1759. (Clap published observations of another comet in the 12 January 1760 issue.)

[80] 14 May 1759.

[81] *Conjectures Upon the Nature and Motion of Meteors,* pp. 11-13.

[82] The article was signed "B.J." One authority speculates that it was John Bevis, a prominent English astronomer.

meteor that had passed over Holland early in 1755, the contributor made a parenthetical observation: "I have this one further remark to add, that supposing this body had been projected parallel to the horizon with an initial velocity of 4,95 miles in a second of time, and that its ignition would not have dissipated or consumed it, nor the atmosphere retarded it, it would have assumed the nature of a sattelite or moon, and revolved around the earth perpetually in a circular orbit." A regular reader of the magazine, Clap doubtless was struck by the account.[83]

The knowledge that at least one European astronomer was conducting research along the same theoretical lines as his own quickened his desire to gain recognition as the author of the theory. In 1756 he set his thoughts down on paper and circulated the manuscript among colonial astronomers.[84] This action came none too soon for three years later John Pringle printed an article in the *Philosophical Transactions* which presented a theory closely paralleling Clap's.[85] The President now intensified his research. On 10 May 1759, a meteor was observed along the coastal area of Massachusetts. Shortly after, Clap, who was en route to Boston on business, travelled through the towns of that locale and interviewed "many people." The information he acquired was sketchy and of a primitive order, offering full testimony to the difficulties of the eighteenth-century astronomer. Residents of Taunton informed him they had heard rumbling noises persisting for two minutes. A man in Roxbury had observed a ball of fire of white transparent brightness about six inches in diameter. A sailor in Boston told him that he had witnessed the phenomena "about a league SE" from Cape Cod. He estimated the meteor to be about "50 to 60 degrees high" and passing from west to east. When it was about one-half mile from him, it burst into a "thousand pieces"; from the standpoint of Clap's theory, this was a shattering observation.

After collating these superficial and oftentimes contradictory observations, Clap drew up a report and forwarded it to John Winthrop, who had issued an appeal in the Boston newspapers for such data.[86] Two years later Winthrop sent an account of the meteor to the Royal Society. The report, which was

[83] The account certainly gave Ezra Stiles a mental jolt when he read it a few years later. Stiles immediately came to an awareness of the similarity in theories. He previously had assumed that Clap was the first astronomer to develop the theory of terrestrial meteors. Now he was not so certain but that Clap had borrowed the idea. Puzzled by the coincidence, he wrote his one-time mentor, informing him of his discovery. He tactfully concluded, however, "I still call it your [hypothesis] though I imagine that Mr. B.J. and yourself both hit upon the same hypothesis by separate and distinct efforts of genius." Clap later sent word to Stiles through a courier that his conclusion of "separate and distinct efforts of genius" was entirely correct. If Stiles had other thoughts on the matter, he refrained from committing them to paper. See the manuscript note by Stiles on the verso of page 1 of John Noyes to Stiles, 5 July 1756; and Stiles to Clap, 19 February 1766; both MSS. are in the Yale University Library.

[84] McKeehan, *Yale Science,* pp. 23-24.

[85] *Phil. Trans.,* 1759, *51:* 259-274. This article was a follow-up to an earlier article in *51:* 218-259. The theories were quite similar with respect to the relative size of the phenomena and their rate of speed. The only major difference involved the attractive body. Clap explicitly affirmed that the earth served as their center of focus. Pringle intimated that the meteors revolved about one central planet, but he failed to designate it. On Pringle, see Dorothea Waley Singer, "Sir John Pringle and His Circle," *Annals of Science,* 1949-1950, *6:* 127-180.

[86] The report, presented in the form of a letter to Winthrop, was published in the *Boston News Letter,* 31 May 1759.

substantially identical with Clap's, was read before the Society and later published in the *Philosophical Transactions,* Winthrop receiving the credit for publication.[87] Clap received not a mention — which leads one to surmise there is truth in Louis McKeehan's statement that the colonial scientists were adept at "stealing the thunder" from each other.[88]

In 1763, after re-working certain points in his theory, Clap sent his manuscript to John Pringle in London.[89] Pringle was to offer it to the Royal Society for a reading and, if merited, publication in the *Philosophical Transactions.* In June of the following year, Peter Collinson, who functioned as a *via media* between the colonial scientists and the Royal Society, wrote Clap that his essay had been received "with Approbation." But he made no mention of publication, confining himself to the editor's timeless platitude that they looked forward to receiving additional accounts from him.[90]

Even before the receipt of Collinson's communication, Clap had returned to his research. Hearing of a meteor that passed over Massachusetts, New York, and eastern Pennsylvania on 10 May 1765, he directed a letter to Ezra Stiles in Newport, Rhode Island. Did he have any information on this "very Remarkable Meteor . . . as big and as bright as the sun. . . . I am in quest of more critical observations?"[91] Through the *Connecticut Courant,* he issued an appeal to "all ministers and all other gentlemen of learning and ingenuity" to correspond with people in the areas in which the meteor had been sighted. Should they travel through those districts, they were to interview people who had witnessed the phenomenon and have them point out the approximate course of the meteor's movement. The men of "learning and ingenuity" were then to compute the point of compass and the altitude at which the meteor had been first and last seen; in the event they did not "well understand degrees," they were to describe it "by its being nearly overhead, or half-way between over-head, and on the horizon, or as high as the sun at three or four hours, or the like." They were then to forward information to Clap or publish it in the newspapers.[92]

When another meteor shot across the sky of southwestern Connecticut on 18 September 1765, Clap gathered together compass and quadrant,[93] hitched

[87] "An Account of a Meteor Seen in New England, And of a Whirlwind Felt in That County: in a Letter to the Rev. Thomas Birch," *Phil. Trans.,* 1761, *52:* 6-16. In another cooperative venture with Winthrop, Clap fared much better. When a meteor passed over southern New England in November 1742, he drew up a report and relayed it to the Harvard scientist. Some twenty-two years later, Winthrop inserted Clap's report, with proper acknowledgment, in a more comprehensive document and sent it off to John Pringle of the Royal Society. Winthrop's paper was read before the Society on 4 June 1764, and published in the *Philosophical Transactions* the same year (*54:* 185-191). This marked the only time that an astronomical report credited to Clap appeared in the famed journal. In 1763, Clap sent to the Society (by way of Pringle) "An Account of Three Meteors Seen in New England," which was read on 7 June 1764. He reported on meteors seen in 1711, 1743, and 1758. The document is housed in the Royal Society.

[88] *Yale Science,* p. 34.

[89] MS, Royal Society.

[90] Dexter, ed., *Stiles' Itineraries,* pp. 452-453.

[91] *Ibid.*

[92] 17 June 1765. By October, Clap had received reports from observers in Coldenham, Salisbury, and Northampton.

[93] Clap utilized the scientific equipment belonging to the college in his personal research. He did own a quadrant, however; see Naphtali Daggett to Mary Wooster and Temperance Pitkin, 14 January 1767, MS, Boston Public Library. One of the orders in President Clap's

horse to chaise, and scurried off on another scientific journey. Although many of the reports he gathered came from people "not skilled in Angles," he was certain that he had tracked down one more "terrestrial" meteor. His final calculations showed it to be about one-half mile in diameter and at a height of thirty-two miles. In October his report appeared in the *Connecticut Gazette.*[94]

Clap made his final appearance in print on 3 March 1766, just three months before he tendered his resignation to the Yale Corporation. He reported on a meteor that had passed over Hartford the previous month. Once again his observations pointed to a terrestrial meteor. He implored observers who had taken compass readings and made a determination of the meteor's altitude and angle of descent to publish their accounts in the newspapers.[95]

The implicit confidence that Clap retained in his theory was not shared by the Royal Society, which probably accounts for the Society's failure to publish his essay. That Clap's theory was based on some tenuous propositions was apparent even to eighteenth-century astronomers. Its weak points have been exposed by a modern authority:

> The orbits had to be very nicely adjusted to miss the earth by about twenty miles every time, and yet the orbits had to change rather rapidly to prevent monotonous repetitions of the observed passages at the same latitudes. (The difference in polar and equatorial radii is so slight as to be unimportant.) Worse yet, the earth wobbles so much about the center of the earth-moon system, nearly 3,000 miles from the center of the earth, that a terrestrial comet which ever got as far away as the moon would have to follow a very remarkable orbit indeed to go on missing the earth so narrowly time after time.[96]

Moreover, as every school boy of the rocket age knows, the earth's gravitational pull at an altitude of ten to twenty miles is much too powerful to allow an object, natural or artificial, to maintain a perpetual orbit. The famous Russian sputnik launched in October 1957, while admittedly a marble compared to Clap's theoretical meteors (the sputnik was only 23″ in diameter and weighed but 184 pounds), was 150 miles from the earth at the lowest point in its orbit (perigee). Yet it disintegrated shortly after two months of earth-circling, a victim of atmospheric friction. Modern astronomical science has also demonstrated that meteorites passing between the earth and the sun are usually shattered by the tidal forces of the sun's gravitational field.

Clap displayed a notable ingeniousness in formulating his theory, but he was conspicuously lacking in what Louis McKeehan aptly calls "physical intuition."[97] His essay on meteors was published posthumously in 1781 at the personal expense of Ezra Stiles.[98] It stands today as a quaint museum piece of early American astronomy. While its general theory is fallacious, it represents one of the earliest efforts by an American (or British-American) to comprehend in a scientific manner a most mysterious aspect of astronomy. Even

Directions to His Children After His Decease, 1757 (photostat, Massachusetts Historical Society) was to return to the college the scientific apparatus he had borrowed.

94 4 October 1765.

95 *Connecticut Courant,* 3 March 1766.

96 McKeehan, *Yale Science,* p. 30.

97 *Ibid.*

98 Stiles presented a copy of the work to the Royal Society in 1784. Dexter, ed., *Literary Diary of Stiles,* II, 119.

today, with hypersensitive chronographs, spectroscopes, spectrographs, and 110-inch telescopes at the disposal of astronomers, a cloud of uncertainty still surrounds the movements of larger meteors.

While Clap's efforts to formulate a theory of meteors can be written off as a failure, his scientific activities as a whole constitute a positive contribution to the history of American science. He was personally responsible for the first systematic teaching of mathematics and the sciences at Yale. By modernizing and revitalizing the curriculum, by procuring apparatus and introducing more up-to-date texts, and by providing for a well-trained tutorial staff, Clap converted the college into an important center of Newtonian science. Moreover, laboratory science at Yale had its origins in the latter portion of Clap's administration, when students assembled in the Library located in the Chapel and "In all Delineations and Calculations, a select Number, with proper Instruments in their Hands, are instructed at a Table."[99] These were the beginnings of a tradition brilliantly perpetuated by the scientific departments of the modern university.

There are more tangible accomplishments to consider. Clap instilled in a generation of Yale youth the main elements of the new science, along with a knowledge of surveying, navigation, and other "practical" subjects. Colonial society felt the effects of his revised curriculum in many subtle ways, from the improvement of agricultural methods to the making of more accurate maps. Further, he was the chief agent in the training of a group of teachers who carried the torch of Newtonian science to another generation of youth. Ezra Stiles and Nehemiah Strong continued the tradition at Yale through the Revolutionary period.[100] Bezaleel Woodward (B.A. 1764), another Clap-trained educator and scientist, had a long and illustrious career as professor of mathematics and natural philosophy at Dartmouth College.[101] Hundreds of ministers trained at Yale under Clap's direction helped promote the scientific revolution in America by dispensing Newtonian theories to the college-bound youth of their parishes. Thus Clap was an important link in the transmission of the Enlightenment to the new world. Quietly, unassumingly, he helped shape the cultural form of American civilization.

Not to be disregarded or minimized are Clap's efforts as a popularizer of astronomical science. His published reports of comets and meteors reached a wide audience in New England. However elementary in content, they served to educate the general public in observational astronomy; and this type of promotional effort played a part in making astronomical research the most widely practiced scientific activity in New England during the post-Revolutionary period.

[99] Clap, *Annals of Yale,* p. 82.

[100] At least on one occasion, Stiles' zeal for astronomy outran discretion. He "Spent five hours incessantly in communicating Instruction to some of the Senior Class in Astronomy, the Calculus and Delineation of a solar Eclipse; calculating the Eclipses of Jupiter Eclipses, the Place of Saturn and the other planets, and the Trajectory and places of Comets both heliocentric and geocentric." Dexter, ed., *Literary Diary of Stiles,* III, 35. Strong, the first occupant of the chair of Professor of Mathematics and Natural Philosophy (created in 1770), clashed with the Corporation over various matters and finally resigned in 1781.

[101] Franklin B. Dexter, *Biographical Sketches of the Graduates of Yale College with Annals of the College History* (New York, 1903), III, 89-92.

Divorced from the scientific program of Yale and judged solely on his personal achievements, he ranks a cut below the top colonial scientists.[102] He lacked the analytical ability of Winthrop, the pragmatic temperament of Franklin. Nor did he possess an architectonic mind. He neither constructed an imposing scientific synthesis nor produced any remarkable discoveries. He was essentially an enlarger or improver, not a creator. He grappled with the advanced problems posed by Cartesian and Newtonian thought, but he did so for the purpose of amplifying rather than creating. His mission was to advance science along rational lines through the doctrines of the "perpetual Dictator," as Cotton Mather reverently alluded to Newton,[103] not to establish a school of scientific thought.

First and foremost, Clap was a Puritan, and like all Puritans he refused to venture at length on the strange sea of thought. If he pushed beyond the known frontiers of contemporary science, as in astronomy, he still maintained limits to his range of investigation and speculation. Try as he might, he could not in conscience follow Pope's exhortation and "mount where science guides."[104]

[102] Had there not been the Revolutionary War, we would have today a larger body of evidence for a more definitive evaluation of Clap's scientific abilities. At that time, a troop of British soldiers plundered the home of Clap's daughter in New Haven, hauling off a box and two trunks, one of which, according to Ezra Stiles, contained Clap's scientific papers. It is reasonable to assume that such an item as his astronomical tables, which the President considered to be a decided improvement over Halley's popular tables and which he had planned to publish (see Clap to William Whiston, footnote 23), was included in this trunk. The British subsequently threw the containers into Long Island Sound, after scattering their contents in the water. President Stiles, enraged by this act of cultural barbarism ("War against Science [Learning]") protested sharply to the British commander. Stiles later walked along the beach and retrieved a few of the documents that washed ashore.

[103] *Manductio ad Ministerium,* p. 50.

[104] The couplet reads:

"Go won'drous creature mount where Science guides
Go, measure earth, weigh air, and state the tides";

Alexander Pope, *An Essay on Man,* edited by Maynard Mack (London and New Haven. 1950 and 1951), p. 56.

John Lining and his Contribution to Early American Science

By Everett Mendelsohn *

CHARLESTON, South Carolina, was one of the liveliest intellectual centers in colonial America. This was in no small measure due to the group of physicians who gathered there during the eighteenth century. The interests of the doctors ranged from medicine to political life, and included literature, botany, and other sciences. In characterizing them William H. Welch has said: "There was no more cultivated and attractive group of medical men in the third quarter of the eighteenth century in America than that in Charleston, S.C. . . . Of these Bull was a pupil of Boerhaave, and Chalmers, Moultrie, Lining, and Garden were trained in Edinburgh. These men were abreast of the knowledge of the day."[1]

Many of the important scientific contributions of this group were made by John Lining. At a time when most of his American colleagues were botanizing, or otherwise exploring the natural history of the colonies, Lining carried out a series of "statical experiments" on human metabolism.[2] However, his contemporaries were more appreciative of his attempt to find a link between weather and disease through a prolonged study of the meteorological phenomena of Charleston. Lining's report on "American Yellow Fever," communicated to the medical society in Edinburgh, was the earliest published description of the disease as it occurred in North America.[3]

* Harvard University. This paper, in a different form, was prepared originally for a seminar on science in America at Harvard University.

1 As quoted in Joseph I. Waring, "Medicine in Charleston, 1750-1775," *Annals Med. Hist.*, n.s., 1935, *7:* 19-26. Brooke Hindle, in *Science in Revolutionary America, 1735-1789* (Chapel Hill, 1956), has pointed out that Bull was the first American to go abroad to study medicine. He went to Leyden in 1734. Hindle points out that "all of the physicians who did make contributions to the advance of natural history in America had experienced a strong Scottish influence; either they were Scots themselves or they had received training in that country" (p. 38). Thomas Dale, James Killpatrick and George Milligen were also members of this group. The accomplishments of these men have been set forth elsewhere. See for instance Robert E. Seibels, "Thomas Dale, M.D., of Charleston, South Carolina," *Annals Med. Hist.*, n.s., 1931, *3:* 50-57. Joseph I. Waring, "James Killpatrick and Smallpox Inoculation," *Annals Med. Hist.*, n.s., 1938, *10:* 301-308.

2 Dr. Alexander Garden, the botanist, was critical of Lining's efforts. In a letter to John Ellis, 6 May 1757, Garden said: "I have engaged Dr. Lining to make some careful experiments on some vegetables, and especially Indigo, to lay before your Society of Arts, London. The worst of it is that the doctor is soon taken with trifling observations, and does not pursue them, else he might be able to do many things . . . ," in J. E. Smith (ed.), *A Selection of the Correspondence of Linnaeus and other Naturalists* (London, 1821), I, 407. Lining was not wholly outside the naturalist tradition. W. M. Smallwood, in

A brief study of Lining's biography and an analysis of his scientific work should provide further understanding of the nature of scientific activity in the colonies in his day. Most of the previous biographical sketches have relied upon the data provided by Dr. David Ramsay in his *History of South Carolina*.[4] These have led to conflicting reports about Lining's medical education and his family.[5]

John Lining was born in the small parish of Walston, Scotland, in April, 1708.[6] He was the eldest son and second of seven children of Rev. Thomas and Ann Hamilton Lining. His father was at the time serving as Presbyterian minister to Walston.[7] Although not well to do, the senior Lining was well respected; he was chosen a Burgess of Glasgow in 1718 and served in a similar capacity for Edinburgh in 1729.[8]

At the age of twenty-two John Lining emigrated to America, settling in Charleston, South Carolina, in 1730. His immediate family biography is provided by excerpts from local parish registers.[9] On 28 June 1739 he married Sarah Hill in the Episcopal Parish of St. Andrew's, Rev. William Guy presiding.[10] Through his marriage Lining entered one of the most prominent families of South Carolina. Sarah Hill was the daughter of Charles and Elizabeth Godfrey Hill. Charles Hill, a wealthy merchant, served as chief justice of the colony, from 1721 to 1724.[11] Dr. Lining and his wife lived at various times

Natural History and the American Mind (New York, 1941), p. 94, reports a manuscript letter (Univ. Edinburgh, Feb. 13, 1759) from Lining to Professor Charles Alston citing the difficulties in sending living plants across the ocean.

[3] John Lining, "A Description of the American Yellow Fever, in a Letter from Dr. John Lining Physician at Charles-Town in South Carolina, to Dr. Robert Whytt Professor of Medicine in the University of Edinburgh," *Essays and Observations, Physical and Literary*, 1756, *2:* 370-395. The account was read before the Edinburgh Philosophical Society, 7 March 1754. A somewhat earlier account of the disease had been prepared by Dr. John Mitchell of Virginia, transmitted to Benjamin Franklin and thence to Dr. Benjamin Rush. However, it was not published until the closing years of the century.

[4] David Ramsay, *The History of South Carolina, from its First Settlement in 1670 to the Year 1808*, 2 vols. (Charleston, 1809). Franklin C. Bing, "John Lining, an Early American Scientist," *Science Monthly*, March, 1928, p. 249. Bing is also author of the note in the *Dictionary of American Biography*, vol. 11, 1933.

[5] See for example the inaccurate reports about Lining in Maurice B. Gordon, *Aesculapius Comes to the Colonies. The Story of the Early Days of Medicine in the Thirteen Original Colonies* (New Jersey, 1949); and Francis R. Packard, *History of Medicine in the United States*, 2 vols. (New York, 1931). Accurate biographical information is included in Robert C. Aldredge, "Weather Observers and Observations of Charleston, South Carolina 1670-1871," *Year Book City of Charleston*, 1940, pp. 190-257. Walter J. Meek, "The Charleston Medical Meteorologists," *Texas Rep. Biol. & Med.*, 1955, *13:* 272-86, has digested the Aldredge material for a talk presented to a local medical society.

[6] Parish Registers of Walston, as noted in Aldredge, *op. cit.*, p. 204.

[7] Hew Scott, *Fasti Ecclesiae Scoticanae. The Succession of Ministers in the Church of Scotland, from the Reformation* (Edinburgh, 1915), I, 263. The report notes that "Thomas Lining, born Lanarkshire; studied at the Univ. of Glasgow; licen. by Presb. of Lanark 16th April 1701; called 14th Feb., and ord. 10th May 1705; died 20th Dec. 1731."

[8] Aldredge, *op. cit.*, p. 204.

[9] A. S. Salley, *Death Notices in the 'South Carolina Gazette,' 1732-1775* (Columbia, S.C., 1917). Also, A. S. Salley, *Register of St. Philip's Parish, Charles Town, South Carolina, 1720-1728* (Charleston, 1904); Mabel L. Webber, "Register of St. Andrews Parish, Berkeley County, South Carolina, 1719-1774," *South Carolina Historical and Geneological Magazine* (hereafter cited as *S.C.H.G.M.*) 1912, *13:* 213-233.

[10] Mabel L. Webber, *loc. cit.*, p. 219.

[11] Anne K. Gregorie and J. Nelson Frierson, *Records of the Court of Chancery of South Carolina, 1671-1779* (Washington, D.C., 1950), p. 436.

in the Hill town house on Broad and King streets in Charleston and at Hillsborough, the family plantation.[12]

A recent popular history of medicine in colonial times notes Lining's marriage to Sarah Hill and remarks that "no children developed from this union."[13] Nothing could be further from the truth. Lining had eleven children in all. Four—Sarah, Mary, Charles Hill and James—died in their youth. Seven others survived Lining's death in 1760, but only one, Major Charles Lining, survived Mrs. Lining's death in 1789.[14]

John Lining died on 21 September 1760. In the next edition of the *South Carolina Gazette* the following note was published:

> On Sunday last died, very much lamented, John Lining, Esq; a gentleman eminent for his application and experience in discovering the causes, nature and cure of the disorders incident to this province where he had practiced physic upwards of twenty years; and who possessed all the good qualifications that could render his loss great, as a physician, husband, father, master, friend, neighbor, companion, &c.[15]

There is no extant record of the type of education John Lining received. It is clear, however, that he did not receive a medical degree from Edinburgh University, nor did he attend the University in any other capacity, for his name fails to appear in the listing of degrees granted or students in attendance.[16] A similar inquiry into the rosters of other Scottish and English universities and the Medical School at Leyden makes it doubtful that Lining received any formal medical training.[17]

The fact that Lining was not a graduate does not imply that he was wholly untrained or a "quack" when he arrived in Charleston in 1730. Like many other physicians of that time, Lining probably received his medical instruction as an apprentice to a practicing doctor. During the colonial period the European-trained physicians were always a minority and some of these, having no more than a single short winter's course at Leyden or Edinburgh, assumed a medical degree on crossing the ocean.[18]

[12] Henry A. M. Smith, "Old Charles Town and its Vicinity . . . With some Adjoining Places in Old St. Andrews Parish," *S.C.H. G.M.*, 1915, *16:* 58. The King Street house is still standing and is marked as the site of Dr. John Lining's medical practice and as the place where he carried out his meteorological observations.

[13] M. B. Gordon, *op. cit.*, p. 413.

[14] A. S. Salley, *Register of St. Philip's Parish*, pp. 83, 85, 91. Also A. S. Salley, *Marriage Notices in the South Carolina and American General Gazette, 1766-1781* (Columbia, S.C., 1914), p. 9, and A. S. Salley, *Marriage Notices in the South Carolina Gazette, 1732-1801* (Albany, 1902). Major Charles Lining fought in the American Revolution and later became a prominent attorney. He maintained the Hill family property until his death in 1834.

[15] A. S. Salley, *Death Notices, South Carolina Gazette*, p. 28.

[16] A perusal of the *Nomina eorum qui gradum Medicinae Doctoris in Academia Jacobi Sexti Scotorum Regis, quae Edinburgi est, adepti sunt* (Edinburgh, 1846), gives no indication of Lining's having defended a thesis. *A Catalogue of the Graduates in the Faculties of Arts, Divinity and Law, of the University of Edinburgh, Since its Foundation*, Edinburgh, 1758.

[17] The registers of Aberdeen, Dublin, Glasgow, St. Andrews, and Leyden have no entry for John Lining. However, W. Innes Addison (ed.), *The Matriculation Albums of the University of Glasgow From 1728 to 1858* (Glasgow, 1913), notes entries for Thomas and James Lining, both sons of Rev. Thomas Lining of Lanarkshire. Thomas, Jr. matriculated 1746, M.A. 1751. James, matriculated 1752, died 1754.

[18] Whitfield J. Bell, Jr., "Medical Practice in Colonial America," *Bull. Hist. Med.*, 1957, *31:* 442-453.

Lining must have gathered some knowledge of the arts of healing, for his first occupation on arriving in the American colonies was as a pharmacist; shortly thereafter he was practicing medicine.[19] The situation was not unambiguous, since many physicians made and sold medicines, and many druggists, wholly lacking in medical training came to be addresed as "Doctor" and practiced healing. That this confused state of affairs existed in Charleston is evident in a letter in the *Gazette*. The correspondent inquired "whether an Apothecary or Chirugeon can be calculated a *graduate* Physician, or a graduate Physician is of Course an Apothecary and Chirugeon, I won't pretend to say."[20] Further indication, however, that Lining was in contact with the medical profession in Edinburgh is seen in the fact that Dr. Robert Whytt, a prominent physician and professor at the University, requested that Lining send him a report on the American yellow fever. This report and another on "The Anthelmintic Virtues of the Root of the Indian Pink," were read before the Philosophical Society in Edinburgh and subsequently published in their journal.[21]

Charleston must have offered a good living for the physician, for during the years 1732 to 1738 the *South Carolina Gazette* lists some thirty-six men who were practicing.[22] Even if some of these men were "quacks" this was still a large number of doctors for a population of under four thousand, of whom more than half were slaves. The practicing physician was probably kept quite busy, for in addition to normal illnesses Charleston was periodically subjected to diseases of epidemic proportions. Malaria, smallpox, yellow fever and dysentery kept the city in constant fear.[23] An epidemic of yellow fever had recently ended when Lining arrived in 1730, and it recurred two years later.

Late in 1737, or early in 1738, Lining was appointed to the position of Parish Doctor and was connected with the newly founded St. Philip's Hospital. The facility had been established through the initiative of local citizens who had petitioned the Governor for permission to raise the money needed to construct a combined hospital and poor house.[24] The building was probably started in 1736 and was completed prior to April 1737 when the vestry of St. Philip's parish announced that a board of five Commissioners had been elected to direct the hospital's activities.[25] Although the hospital was commonly referred to as the "poor house," the fact that physicians were connected with it and that the vestry recommended patients for treatment makes it quite certain that at least one of its major functions was the dispensing of medical

19 Aldredge, *op. cit.*, p. 205, cites the *South Carolina Gazette* of 4 August 1733, which carried an advertisement by John Lining who was selling drugs, etc. Hennig Cohen, *"The South Carolina Gazette," 1732-1775* (Columbia, S.C., 1953), p. 44, records the first note in the *Gazette*, which mentions Lining as a practicing physician, of 24 November 1738. Whitfield Bell, *loc. cit.*, p. 444, notes that it was not at all uncommon for a man to enter medical practice after training or experience as an apothecary.

20 *South Carolina Gazette*, 12 June 1755, as quoted in Hennig Cohen, *op. cit.*, p. 43.

21 John Lining, "American Yellow Fever," *Essays and Observations*, 1756, *2:* 370, and *1:* 386-389.

22 St. Julien R. Childs, "Notes on the History of Public Health in South Carolina, 1670-1800," *Proc. South Carolina Hist. Assoc.*, 1932, p. 21.

23 Yellow fever, for instance, hit Charleston severely in 1699, 1706, 1728, 1732, 1739, 1745, 1748. Of the colonial cities Charleston suffered most heavily from epidemic diseases. John Duffy, *Epidemics in Colonial America* (Baton Rouge, 1953).

24 Joseph I. Waring, "St. Philip's Hospital in Charles-Town in Carolina: Medical Care of the Poor in Colonial Times," *Annals Med. Hist.*, n.s., 1932, *4:* 285.

25 St. Julien R. Childs, *loc. cit.*, p. 20.

care. The vestry records contain an itemized bill from Dr. Lining which mentions his services during a smallpox epidemic of 1738. By 1740, however, Lining was no longer serving with the hospital.

The colonial physician was deeply involved in the affairs of his community. While still new in Charleston Lining joined with a group of other Scottish immigrants in founding the St. Andrews Society. He also became a Mason, and the *Gazette* of December, 1738, reported that he had been elected Senior Warden of Soloman's Lodge.[26]

It is probable that John Lining was among the group of men that met in 1748 to plan an association that would raise funds to purchase pamphlets, magazines and books from Great Britain. In its second year this group, which became known as the Charles-Town Library Society, elected Lining as their president. Of the one hundred twenty-nine members listed for April, 1750, fifteen were physicians, indicating some interest on the part of these professional men in the circulation of literature.[27]

The General Assembly gave further recognition of Lining's growing reputation. When it passed an act in 1747 requiring inspection of boats coming into the harbor Dr. John Lining was the first named of the panel of physicians appointed to carry out the act.[28] He was also called on by the courts to act as an arbitrator in disputes, especially those concerning medical affairs. A document preserved by the Historical Association contains the Heron-Lining Decision, 1751: "The decision rendered to the Court of Common Pleas by Alexander Heron and John Lining, arbiters in a dispute between David Oliphant ('practitioner in physic in Charles Town and surgeon to the three Independent Companies in South Carolina') and Paschal Nelson (commander of one of the companies) over medicine furnished by the former."[29]

One interesting aspect of Lining's medical practice is that for part of the time he shared his office with Dr. Lionel Chalmers. Although there is no indication when this partnership commenced, its termination was announced in an advertisement in the *Gazette* of 24 February 1754. It read:

> We desire the Favour of all Persons indebted to us, to pay their respective debts on or before the first of April next; and as our co-partnership terminates the last day of this month it is hoped due regard will be had to this nottice. Feb 1st, 1754. Lining & Chalmers.[30]

Lining's scientific work may have influenced his partner, for Dr. Chalmers held ideas similar to those of Lining regarding the importance of weather in disease. Chalmers began a series of meteorological observations at about the time Lining's ended. They were similar to Lining's, but did not show the same meticulousness of recording. The majority of the observations were recorded in the two volumes Chalmers wrote giving *An Account of the Weather and*

[26] *South Carolina Gazette,* 28 December 1738.

[27] Arthur Mazyck (ed.), "Original Rules and Members of the Charleston Library Society," *S.C.H.G.M.,* 1922, *23:* 163 ff.

[28] Thomas Cooper (ed.), *The Statutes at Large of South Carolina,* Edited under Authority of the Legislature, vol. 3, 1716-1752 (Columbia, S. C., 1838), p. 695.

[29] Helen G. McCormack, "A Provisional Guide to Manuscripts in the South Carolina Historical Society," *S.C.H.G.M.,* 1944, *45:* 112.

[30] *South Carolina Gazette,* 12 February 1754, as quoted in Aldredge, *op. cit.,* p. 215.

Diseases of South Carolina.[31] Chalmers made direct reference to Lining's work and utilized the latter's meteorological tables. However, Chalmers failed to describe his own instruments and did not indicate their placement during observations, thus making his work less valuable than Lining's.

In his *Essay on Fevers* Chalmers made intensive use of the data gathered by Lining in the "statical experiments."[32] Commenting on his relationship to Lining he said that some time after the conclusion of the "statical experiments" he "became . . . so closely connected with him, both by mutual interest and friendship, as gave him an opportunity to be fully informed of all he desired to know, relating to the perspiration and other natural discharges in each season."[33] Chalmers utilized this information to challenge the Sanctorian notions about the connection between perspiration and illness.

Lining may have found that partnership in medical practice provided him with the free time needed for his scientific pursuits, for he no sooner dissolved one tie when he entered upon another. The *Gazette* of 26 February 1754 announced that "the Partnership between Lining and Chalmers expires next Thursday; and on the Day following, being the First of March, the Partnership between Lining and Oliphant commences."[34] Dr. Oliphant, as mentioned above, had been a physician serving three companies of the King's troops in South Carolina. It was about this time that Lining gave up medical practice and retired to Hillsborough, the family plantation.[35]

Another incident involving the members of the medical profession of early Charleston concerned a group called the Faculty of Physic. This group, of which Dr. John Moultrie was president, was dedicated to "the better Support of the Dignity, the Privileges, and Emoluments" of the "Humane Art."[36] The *Gazette,* after announcing their meeting, pointed out that the members, "considering that they are often called out under the greatest Inclemencies of the Weather, sometimes merely to gratify the Patient, and sometimes when no Medicines are required . . . and likewise that they are often slowly and seldom sufficiently paid, for their sollicitous care . . . therefore have unanimously resolved, that after the Tenth Instant, they will give no further attendance without a reasonable Fee, paid at the first visit, and at every other visit, during the Course of their Attendance."[37] The community reacted with caustic comment, and one wit penned a poem:

> The Doctors in Charles Town have lately agreed
> Not to visit their Patients until they're new Fee'd,
> From whence this Suggestion most truely arises,
> What Fee must be paid? And what the New Price is?[38]

[31] Lionel Chalmers, *An Account of the Weather and Diseases of South Carolina* (London, 1776), 2 vols.

[32] Lionel Chalmers, *An Essay on Fevers: More Particularly those of the Common Continued and Inflammatory Kinds* . . . (London, 1768), passim.

[33] *Ibid.,* p. iii.

[34] *South Carolina Gazette,* 26 February 1754, as quoted in Aldredge, *op. cit.,* p. 215.

[35] Writing in 1757 Dr. Alexander Garden was able to say of Lining: "he has turned a planter now altogether, and has quite done with practice." Garden to Ellis, 6 May 1757. J. E. Smith (ed.), *op. cit.,* I, 407.

[36] Joseph I. Waring, "An Incident in Early South Carolina Medicine," *Annals Med. Hist.,* n.s., 1929, *1:* 608.

[37] *South Carolina Gazette,* 5 June 1755, as quoted by Waring, *op. cit.,* p. 608.

[38] *Ibid.,* p. 609.

The "Faculty of Physic" was one of the earliest professional societies formed in North America.[39] It would be interesting to know whether the concern of the members extended to the problem of regulating standards of practice as well as remuneration.

2

Among the early medical practitioners in Charleston there were several who supplemented their healing efforts with experimentation, writing, translating medical treatises, and research and observation in various fields of science.

Dr. Thomas Dale, the oldest of this group of Carolina medical men, undertook the translating of several medical texts. Two were by Dr. John Freind, the physician, chemist and disciple of Newton. Freind's approach to medicine and chemistry put him in that group of scientists who hoped to interpret all phenomena in a mechanical way.[40] Through these efforts Dale made several important medical texts available to the Charleston profession and the English-speaking medical world as a whole. Dr. Dale later became involved in a pamphlet battle with Dr. Killpatrick, an advocate of inoculation against smallpox.

Dr. Lionel Chalmers included a number of meteorological observations in his *Account of the Weather and Diseases of South Carolina.* He also attempted to discover the causes of, and suggested treatment for, fevers. Dr. George Milligen, relying heavily on the work of Chalmers and Lining, sought the cause of fevers in the weather and included his investigations of this question in his *Short Description of the Province of South Carolina: with an Account of the Air, Weather and Diseases at Charles-Town.*[41] Alexander Garden is probably the best known of the physicians of colonial Charleston. His botanical collecting kept him in contact with the lively circle of European and colonial natural historians. These activities secured for him election as a Fellow of the Royal Societies at London, at Edinburgh and at Upsala.

Dr. John Lining was very much a part of this company of men who by their efforts hoped to advance science and knowledge in general. It is interesting to note that Lining was one of the very few among the active physicians of Charleston or other colonial centers who turned his scientific interests toward experimentation. Perhaps it is because he stood on the fringes of the prevalent collecting and naturalist tradition that he is often overlooked.

Although scattered attempts to collect temperature and rainfall data were made in England as early as 1664, few prolonged observations were made until about 1723.[42] In that year Dr. James Jurin, then secretary of the Royal So-

[39] Brook Hindle, *op. cit.*, p. 61; also Whitfield Bell, *loc. cit.*, pp. 452-453.

[40] See. *D. N. B.* for John Freind (1675-1728). The English renditions of his Latin texts were: *Emmenologia: Written in Latin by the late Learned Dr. John Freind,* translated into English by Thomas Dale, M.D. (London, 1729) and *Nine Commentaries Upon Fevers: and Two Epistles Concerning the Small-Pox, Addressed to Dr. Mead,* Translated into English by Thomas Dale, M.D. (London, 1730). Dale also translated a text by Joost van Lom, *A Treatise of Continual Fevers: In Four Parts* (London, 1732).

[41] Originally published London, 1770. Recently reproduced in Chapman J. Milling (ed.), *Colonial South Carolina: Two Contemporary Descriptions by Governor James Glen and Dr. George Milligen-Johnston* (Columbia, S.C., 1951), p. 21.

[42] The early recordings were made by the versatile Robert Hooke. Background material in Gordon Manley, "The Weather and Diseases; Some 18th Century Contributions to Observational Meteorology," *Notes & Records Roy. Soc.*, London, 1952, *10:* 300-307,

ciety, prepared a paper for the Philosophical Transactions in which he urged the making of continuous weather observations.[43] These, he emphasized, should be made in some uniform manner. He recommended using the Hauksbee thermometer. This instrument, which was similar to those described as "Royal Society thermometers," was graduated from zero to ninety with the freezing point being about 76 to 79 degrees; the more temperate heats were lower in number.[44] Lining in his meteorological observations claimed to have used an instrument similar to this one as well as the more practical thermometer by Fahrenheit.[45]

Jurin, who had studied medicine with Boerhaave at Leyden, believed then, as Lining did later, that there was a definite correlation between weather and disease. It is, therefore, not surprising that many who responded to Jurin's appeal for accurate meteorological observations were physicians.[46] Lining's records, beginning in 1737, mark him as one of the earliest keepers of weather data on either side of the ocean, and certainly in the British North American colonies. Paul Dudley, the Chief Justice of Massachusetts, had made non-instrumental observations in 1729-1730, and although they were communicated to the Royal Society, they were never published.[47] Lining's were the first instrumental records kept, and possess, in a remarkable degree, many of the refinements of later years.[48]

The method urged by Jurin consisted of using the Hauksbee thermometer and as "the most convenient site . . . a room facing north where a fire never burns, or as rarely as possible."[49] The temperature records thus made, proved hard to use. It was not until the early 1750's that notable developments were made in temperature measurement. During that decade the Fahrenheit thermometer came into wider use, and, in response to critical letters in contemporary journals, it was realized that outdoor observations in the shade would be of greater significance.[50] If this was actually the case, Lining's techniques and instrumentation did indeed possess refinements of later years. As early as 1740 in his report on the "statical Experiments" he recorded that "the Thermometer is *Fahrenheit's*." He does go on to add "the other Thermometer is made by *Thomas Heath*, in *London;* and is divided into 90 equal Parts; 65 is the freezing point, and 49 temperate: I suspect it to be the same with *Hauksby's* and have called it so in the Tables."[51] Lining further reported that his instruments were kept outside the northeast window of his house and were protected from the sun and rain by a specially constructed box.

That the innovations made by Lining were duly appreciated by his contemporaries is illustrated by one of those "critical contemporary letters" mentioned above. In a communication to the *Gentleman's Magazine* in 1750 entitled "An Essay on some of the Uses of Natural History, with a special View to certain

---

and Alfred J. Henry, "Early Individual Observers in the United States," *U.S. Weather Bureau Bull.*, 11, pp. 291-302.

[43] *Phil. Trans. Roy. Soc.*, London, 1723, *32:* 422.

[44] Manley, *loc. cit.*, pp. 301, 304.

[45] John Lining, "Statical Experiments," *Phil. Trans. Roy. Soc.*, London, 1742/3, *42:* 497.

[46] For a list of English physicians who kept weather records see Manley, *loc. cit.*, passim.

[47] Henry, *loc. cit.*, p. 295.

[48] *Ibid.*, p. 295.

[49] Manley, *loc. cit.*, p. 301.

[50] *Ibid.*, p. 302.

[51] John Lining, "Statical Experiments," p. 497.

improvements in the British Colonies of North America," the correspondent makes a plea for more "observations on the variations of the barometer, thermometer, quantities and times of rain . . . etc."[52] He suggests that many defects in the observations made were due to a dissimilarity of instruments. "This defect," he adds, "it is now hoped, will be remedied in time by a general use of *Fahrenheit's* mercurial thermometer in all observations on the degrees of heat and cold in different places."

This anonymous contemporary critic of weather observations, after pointing to the faults and suggesting remedies, notes that "the indefatigable Dr. Lining, of *Charles Town* in *South Carolina,* has set a good example of this kind: An abstract of whose diaries has been presented to the publick in the *Philosophical Transactions:* These indeed were originally formed with a view to the practice of physick in that province; but may serve, as far as they go, to other purpose, for the sake where of it were to be wished that ingenious persons in the several colonies of *North America,* would pursue some observations of the like kind." This same correspondent used Lining's table in a comparison of the temperature of Carolina with that of other areas within or near the tropics.

Dr. John Lining's first weather reports were communicated as part of the "statical experiments" to Dr. James Jurin dated from Charleston, 22 January, 1740/1. They were not read before the Royal Society until 19 May 1743. The meteorological data were collected with the Fahrenheit and Hauksbee thermometers noted above, and also a barometer which is described as "a common portable one; the Diameter of its Bore is about ⅕ of an Inch."[53] It is probable that Lining was utilizing an instrument similar to the one first described by Rowning.[54] It is noted as an improvement "in the common Barometer, whereby it is rendered *portable.* The Tube containing the Mercury, instead of having its lower End immerged in a Vessel of that Fluid, has it tied up in a Leathern Bag, not quite full of Mercury. And though the external Air cannot get into the Bag to suspend the Mercury in the Tube, by pressing on its Surface . . . yet it has the same Effect by pressing on the Outside of the Bag, which being pliant, yields to the Pressure, and keeps the Mercury suspended in the Tube at its proper Height."[55]

Lining also made use of a hygroscope or hygrometer which he described as "a whip-cord, prepared after the same Manner as that of the Society's in *Edinburgh;* the Difference betwixt its greatest and least Length, by their Manner of Preparation, I found to be Five Inches; for which I made an Index Five Inches long, and divided it into 100 equal Parts, the First of which is the Hygroscope's greatest Length."[56] Records were also provided for the depth of rainfall, thunder (giving indication of its violence) and the force of the wind. The last two of which, he said, "I am obliged to judge by My Senses." Lining's failure to utilize an instrument for measuring the force of the wind may well

[52] A.B.C.D., "An Essay on Some of the Uses of Natural History, with a Special View to Certain Improvements in the British Colonies of North America," *Gentleman's Magazine,* 1750, *20:* 493.

[53] John Lining, *"Statical Experiments,"* p. 497.

[54] John Rowning, *A Compendious System of Natural Philosophy: with Notes, Containing the Mathematical Demonstrations, and Some Occasional Remarks,* 7th ed., 2 vols. (London, 1772).

[55] *Ibid.,* I, 113. A similar description of the "common portable" barometer is given by Charles Hutton, *A Mathematical and Philosophical Dictionary* (London, 1796), I, 189.

have been due to his reliance upon advice given by the Edinburgh society which he mentions, for in the scheme for meteorological observations with which they preface their *Medical Essays* they make no mention of a machine for measuring the force of the wind.[57]

Lining continued to provide meteorological data through 1753. His reports found their way into the *South Carolina Gazette*, and the *Gentleman's Magazine*, as well as the *Philosophical Transactions of the Royal Society* of London.[58]

The tables prepared by Lining were used in yet another way. In his Governor's Report, James Glen included a section on "the Nature of the Climate; uncommon Extreams of Heat and Cold; Tabular Accounts of the highest and lowest Altitudes of the Barometer, of the Depths of Rain, and of the Winds Direction; various Observations relating to Heat, Cold, Vegetation . . . &c."[59] In this section of the report, actually completed in 1749, Glen remarks:

> I had for some Time kept a Diary of the Weather, to please myself only; but having met with a Gentleman here, who is curious in my own Way, and who hath done it with more Accuracy, than the little Portions of Time stolen from the Duties of my Station, would permit me to do; I shall here give you his Tables, which are the Result of Four Years of Barometrical Observations taken Twice a Day, *viz.* at Noon and Night; and of Four Years Thermometrical Observations by *Fahrenheit's* Thermometer; and also, his Account of the Depths of Rain which have fallen in *Charles Town* within each Month and Year for Eleven Years past; together with a Table of the winds.[60]

It is the barometrical tables, beginning in April 1737, which were the earliest recorded meteorological observations made by Lining. After presenting the tables Glen directly acknowledged their source. "These Observations and Tables," he said, "were made and formed by a very curious Gentleman, one Doctor *L-n-g;* and to them I shall add a few other Observations."[61]

It is quite probable that the observations by Lining represent the earliest sustained effort at meteorological reporting done with instruments in the American colonies. Paul Dudley's observations were conducted without the benefit of instruments and were never published. The excellent series of meteorological reports by John Winthrop, Hollis Professor of Mathematics and Natural Philosophy at Harvard, were not begun until 1742, although once started they

[56] John Lining, "Statical Experiments," p. 498. Charles Hutton, *op. cit.,* I, 612, gives a description of several early "whipcord" hygroscopes. The principle is similar in all cases. The one briefly mentioned by Lining probably consisted of a whipcord or cat-gut stretched over one or more pulleys; secured at one end and with a weight fastened at the other, its length would vary with changing atmospheric moisture. An indicator attached at the weighted end served to show change of length when positioned alongside a scale.

[57] This omission is pointed out by a contemporary English weather observer; Roger Pickering, "A Scheme of a Diary of the Weather; together with Draughts and Descriptions of Machines Subservient thereunto," *Phil. Trans. Roy. Soc.,* London, 1744/5, *43:* 2. Pickering's notes about observations and instruments utilized are interesting for comparison. Lining, with the exception of the anemometer, was using instruments and concepts as up-to-date as those of Pickering.

[58] Claude Bernard, *Leçons sur la chaleur animale* (Paris, 1876), p. 336, cites Lining's meteorological observations in a discussion of life in a warm climate.

[59] From James Glen, "A Description of South Carolina; Containing Many curious and interesting Particulars relating to the Civil, Natural and Commercial History of that Colony . . ., London, 1761," in C. J. Milling, *op. cit.*

[60] *Ibid.,* p. 20.

[61] *Ibid.,* p. 36.

were continued for some 20 years.[62] Lining's techniques and data compared favorably with those of Winthrop. The latter, it is interesting to note, utilized the somewhat unsatisfactory Hauksbee thermometer until 1759 when he changed to the Fahrenheit instrument.[63]

Lining's fifteen years of observation of the various aspects of weather and climate make up his most sustained scientific work. In his one year of "statical experimentation" he showed his greatest originality. His first written efforts included meteorological data, but the bulk of the report was concerned with "Statical Experiments made several times a Day . . . for one whole Year."[64] In attempting to make clear the reasons for the compilation of the tables accompanying his letter, Lining offered the following explanation:

> What first induced me to enter upon this Course, was, that I might experimentally discover the influences of our different seasons upon the Human Body; by which I might arrive at some more certain knowledge of the causes of our epidemic Diseases, which as regularly return at their stated Seasons, as a good Clock strikes Twelve when the Sun is in the Meridian; and there for must proceed from some general Cause operating uniformly in the returning different Seasons.[65]

In the quantitative data he collected Lining hoped to find the real link between disease and weather, especially the epidemic diseases which returned each year with such great regularity.

Investigations of this sort were not new. Attempts to measure effectively the intake and excrement of the body date back to the early seventeenth century when Sanctorius carried out his famous experiments in the "weighing chair." This chair, which was suspended from one end of a steelyard balance, was used by Sanctorius for recording variations in his body weight under all sorts of conditions. He measured his food intake and all the possible types of body excreta, calculating the sensible and insensible perspiration. His conclusions were presented in a series of about five hundred aphorisms.[66]

A similar series of observations were made by James Keill, and the results were printed in part along with an English translation of those of Sanctorius. These also were known to Lining. Keill's efforts were occasioned by the belief that the climatic differences between Italy, the site of Sanctorius' observations, and England, the site of his own, required new computations. John Quincy's introduction to these observations relates that "Dr. James Keil of Northampton, a judicious and curious Person, hath been at the Pains to make the same Tryal at the Place where he lived, so far at least as was necessary and conducive to regulate and adjust Sanctorius's Calculations to our own Country."[68] Although it is further claimed that Keill kept records "throughout a whole Year, as observed every Day at the several Hours of the Day, with the Cir-

[62] Henry, *loc. cit.*, pp. 296 ff.

[63] Frederick Brasch, "John Winthrop (1714-1779), America's First Astronomer and the Science of his Period," *Pubs. Astr. Soc. Pacific*, 1916, *165:* 8-9.

[64] John Lining, "Statical Experiments," p. 491.

[65] *Ibid.*, p. 492.

[66] Sanctorius, *Medicina Statica: Being the Aphorisms of Sanctorius, Translated into English, with Large Explanations. To which is Added Dr. Keil's Medicina Statica Britannia . . . also Medico-physical Essays . . . by John Quincy, M.D.* (London, 1723).

[67] John Lining, "Statical Experiments," p. 492.

cumstances of Living, and Conditions of Health that whole Time, as also the Temperature and Weight of the Air, as it appeared by the Thermometer and Barometer," Lining was not satisfied with the results.[69]

"Keil," observed Lining, "indeed has obliged the World with his *Statical Experiments,* but these his extensive Practice made less perfect than he could have wished, having many deficient Days, and he seldom gives the diurnal Perspiration." In fact Lining felt that his own were the first experiments of the sort which presented consistent, accurate data. He reports that "the Method I have observed in the Tables is this: I weigh myself twice every Day, once in the Morning immediately after I rise, and again before I go to Bed at Night." He also measured, at regular intervals, the quantity of urine, stool and perspiration. "The Number of Pulses" was taken "in the Morning, and immediately before . . . Bed at Night."[70]

Lining was not wholly satisfied with his tables, for he wished to "have added an Analysis of a little of my own Blood and Urine, in every Month, with Blood's Cohesion." He was unable to do so because he did not have the necessary instruments. Later when instruments were available he found that to measure the specific gravity of the blood, with the "hydrostatic balance" at his disposal would require more blood than could safely be taken.[71]

It would seem that Lining's "Statical Experiments" did not receive immediate attention, for in April, 1741, some four months after his first letter, Lining forwarded another letter to James Jurin in London.[72] The communication from Lining that brought the data of the "Statical Experiments" up through the one year promised was not mailed from Charleston until 29 January 1743.[73] In this report Lining collected some of the data into a single table in order to "point out the physical Principles, from whence we may account for the Production of these epidemic Diseases of the different seasons, which are not infectious."[74]

With regard to the diseases of the season Lining asks, "are not these the Effects of different Constitutions of the Air on human Bodies?" He continues, "are not the Increments and Decrements of the sensible and insensible Excretions, Regard at the same time being had to the Quantity and Quality of the *Inqesta,* and to Exercise, &c. the only *Index* of the changes produced in the human Constitution, by the Vicissitudes of the Weather? That indeed was the only View I had in going thro' these troublesome Experiments."[75]

Just how much use was ever made of the metabolic data collected, or how far Lining's work went to inspire similar efforts on the part of others, is not known. It is evident that the "Statical Experiments" did not immediately set other scientists or physicians in the American colonies to following their example. These studies completed by Lining were, however, in an important

[68] James Keill, included in Sanctorius, *op. cit.*, p. 321.

[69] *Ibid*, p. 322.

[70] John Lining, "Statical Experiments," pp. 492-493.

[71] John Lining, "A Letter . . . Serving to Accompany Some Additions to His Statical Experiments," *Phil. Trans. Roy. Soc.*, London, 1744/5, *43:* 318.

[72] This second letter was read with the first at a meeting of the Royal Society. It was published with it, the pagination being consecutive.

[73] John Lining, "A Letter . . . Serving to Accompany Some Additions to His Statical Experiments," *Phil. Trans. Roy. Soc.*, London, 1744/5, *43:* 318.

[74] *Ibid.*, p. 319.

[75] *Ibid.*, p. 319.

tradition. The iatro-mathematical experimentation upon which Lining's work was modelled led very naturally into the more sophisticated studies of perspiration carried out by Cruikshank and the classic experiments on the human metabolism done by Lavoisier and Séguin. Lining's own efforts stand at the turning point. They are more careful and more cognizant of the needs of accurate measure than were those of Sanctorius and Keill, but they came too early and were too limited in scope to be considered predecessors of the experiments of Séguin and Lavoisier. In their attempt to link disease directly to changes in the weather Lining's efforts were naive, as all pre-germ theory ideas must be. Taken at their face value, however, the "Statical Experiments" secure for Lining a place among the ingenious experimenters of colonial America.

Lining's efforts did not go entirely unnoticed. At least one contemporary physician reporting on colonial conditions recognized the importance of the material collected in the "Statical Experiments." Dr. George Milligen in his *Short Description of the Province of South Carolina* gives his readers a full account of the work done by "the late ingenious Dr. Lining."[76] After summarizing the several tables Milligen concludes: "From Dr. *Lining's* Experiments we may likewise see the Reason why People generally look better, fresher, and fuller in the Winter," and also "why People in the Fall are weakest and faintest."[77] He finds all traceable to the quantity and balance of perspiration and evacuation.

Little more has been said about the "Statical Experiments." Most notice has been given to the more comprehensive, and perhaps more immediately important, meteorological observations. These gave an accurate picture of early weather conditions in Carolina and as such were important to later weather observers. To the historian these observations show that Lining was in contact with English scientific work. The "Statical Experiments," however, indicate that Lining was not only keeping up with scientific progress but was making some effort to find answers to the perplexing problems of weather and its influence on metabolism and disease.

Two other interests of Lining's deserve to be mentioned in this survey of his scientific activities. The first of these was his report on yellow fever. Although Lining's description of the disease, as included in a letter to Dr. Robert Whytt, was not the first written account of the disease, it was the earliest published in English and certainly the first communicated from the colonies to a medical body abroad.[78] From the covering letter it appears that Dr. Whytt had been anxious for Lining to prepare such a report, for Lining said: "In obedience to your desire, I have sent you the history of the yellow fever as it appeared in the year 1748, which, as far as I can remember, agreed in its symptoms with the same disease, when it visited this town in former years."[79]

The report, which covers some twenty-five pages, is primarily concerned with providing an accurate description of the symptoms. In an effort to uncover

[76] George Milligen, "A Short Description of the Province of South-Carolina, with an account of the Air, Weather, and Diseases, at Charles-Town, Written in the Year 1763" (London, 1770), included in C. J. Milligen, *op. cit.*, p. 111.

[77] *Ibid.*, pp. 157-158.

[78] *Essays and Observations, Physical and Literary, Read Before a Society in Edinburgh and Published by Them*, 1756, *2*: 370. See note 3, above.

[79] *Ibid.*, p. 370.

some of these Lining conducted several autopsies, or dissections, as he called them. He regretted that he "could not give a fuller account of the dissections of those who died of this disease, having unfortunately lost my notes taken from those dissections."[80] It is unfortunate that a further description was not available, for it could have provided an interesting document from which a greater appreciation of Lining's medical skills might have been gained.

The report is otherwise not outstanding, although Lining does conclude that "this fever does not seem to take its origin from any particular constitution of the weather."[81] This insight was gained, he said, from observation of the disease over a period of twenty-five years; in only four of these had it reached epidemic proportions. Lacking the modern knowledge of insect transmission, Lining further concludes "that this is really an infectious disease." This seemed apparent to him because the nurses often came down with the disease and also "as soon as it appeared in town, it soon invaded newcomers, those who never had the disease before, and country-people when they came to town, while those who remained in the country escaped it."[82] The disease was equally baffling to other contemporary observers.[83]

The second of the minor scientific interests of John Lining was one he shared with many of his contemporaries. His interest in electricity and lightning dates from the period shortly after Benjamin Franklin began his experiments in 1752. Lining's own experiments, all confined to the seven years prior to his death in 1760, were first reported in a letter he sent to the *Gentleman's Magazine* in May, 1753.[84]

In this letter Lining relates that "I have several times this season when there was an appearance of a thunder storm, succeeded in making Mr. *Franklin's* experiment with a kite, for drawing the lightning from the clouds, and last *Monday* repeated the same with remarkable success before many spectators."[85] In another communication published in the *Philosophical Transactions* Lining answers several queries from Charles Pinckney (Carolina colonial agent in London) dealing with electrical phenomena.[86]

These investigations conducted by Lining, although of interest to a biographer of the Charleston physician, are otherwise important only as they are viewed as a part of the developing knowledge of and widespread interest in electricity. In was this interest, however, which brought Lining into contact with Benjamin Franklin. In an exchange of several letters Franklin related the results of experiments in electricity and also told of his attempts at rapid cooling by evaporization.[87]

Lining's many scientific efforts mirror the interests of the eighteenth-century

80 *Ibid.*, p. 371.

81 *Ibid.*, p. 372.

82 *Ibid.*, p. 373.

83 For discussion, see John Duffy, "Yellow Fever in Colonial Charleston," *S.C.H.G.M.*, 1951, *52:* 189-197.

84 John Lining, "Extract of a Letter from the Ingenious Dr. John Lining of Charles Town, South Carolina," *Gentleman's Magazine,* 1753, *23:* 431.

85 *Ibid.*, p. 431.

86 John Lining, "Extract of a Letter From John Lining, M.D., of Charles Town, in South Carolina, to Charles Pinckney, Esq; in London: With Answers to Several Queries Sent to Him Concerning His Experiment of Electricity With a Kite," *Phil. Trans. Roy. Soc.*, London, 1754, *48:* 757.

87 Benjamin Franklin, *Experiments and Observations on Electricity, Made at Philadelphia in America,* 4th ed. (London, 1769), pp. 363-367. The letter on evaporation is probably the earliest account of the relation of evaporation and rapid cooling.

inquirer and show that even the lesser figure in the history of science can make a unique contribution. Most of the active physicians in colonial America turned toward natural history for their scientific endeavors. A few, Lining among them, tried to make their contributions through experimentation. None achieved the success of their British and continental counterparts. All, however, served to enrich the developing culture of the North American colonies.

Appendix

The Published Writings of Dr. John Lining of Charleston, South Carolina

"Extracts of Two Letters from Dr. John Lining, Physician at Charles-Town in South Carolina, to James Jurin, M.D., F.R.S. Giving an Account of Statical Experiments Made Several Times a Day Upon Himself, For One Whole Year, Accompanied With Meteorological Observations; To which are Subjoined Six General Tables, Deduced From the Whole Year's Course." *Philosophical Transactions Roy. Soc., London,* 1742-1743, *42:* 491.

"A Letter From Dr. John Lining, Physician at Charles-Town in South Carolina, to James Jurin, M.D., F.R.S. serving to Accompany Some Additions to His Statical Experiments." *Philosophical Transactions Roy. Soc., London,* 1744-1745, *43:* 318.

"A Letter From Dr. John Lining to C. Mortimer, M.D. Sec. R.S. Concerning the Weather in South-Carolina; With Abstracts of the Tables of His Meteorological Observations in Charles-Town." *Philosophical Transactions Roy. Soc. London,* 1748, *45:* 336.

"A Letter From John Lining, M.D. of Charles-Town, South Carolina, to the Rev. Thomas Birch, D.D. Secr. R. S. Concerning the Quantity of Rain Fallen There From January 1738, to December 1752." *Philosophical Transactions Roy. Soc. London,* 1753, *48,* pt. 1: 284.

"From the Carolina *Gazette.* A Comparison of the Late Excessively Dry Weather With that of Some Preceding Years." *Gentleman's Magazine,* 1753, *23:* 121.

"Extract of a Letter From the Ingenious Dr. John Lining of Charles Town, South Carolina." *Gentleman's Magazine,* 1753, *23:* 431.

"Extract of a Letter From John Lining, M.D. of Charles Town in South Carolina, to Charles Pinckney, Esq; in London: With His Answers to Several Queries Sent to Him Concerning His Experiment of Electricity With a Kite." *Philosophical Transactions Roy. Soc. London,* 1754, *48,* pt. 2: 757.

"Of the anthelmintic Virtues of the Root of the Indian Pink, being Part of a Letter from Dr. John Lining Physician at Charles-town in South Carolina, to Dr. Robert Whytt, Professor of Medicine in the University of Edinburgh." *Essays and Observations, Physical and Literary,* 1754, *1:* 386-389.

"A Description of the American Yellow Fever, in a Letter from Dr. John Lining Physician at Charles-town in South Carolina, to Dr. Robert Whytt Professor of Medicine in the University of Edinburgh." *Essays and Observations, Physical and Literary,* 1756, *2:* 370-395.

A Description of the American Yellow Fever, Which Prevailed at Charleston . . . in . . . 1748, Philadelphia, 1799.

Gassendi in America

By Mel Gorman *

WHEN seventeenth-century Englishmen planned to emigrate to America, their preparation involved the gathering of a minimum amount of equipment necessary to gain a foothold on the forbidding and unknown shore and to survive during a period of grim struggle against the wilderness. One should think that the mere contemplation of such an undertaking would eliminate from the mind of the prospective colonist any thought of books, save perhaps a utilitarian handbook or two. Although the first settlers left their homeland for various economic, political, religious, and social motives, they were deeply infused with the English middle-class concept of self-improvement through purposeful reading. Furthermore, they were possessed of an evangelical cultural mission, and were determined that their minds would not be stultified by any barbarism which might be engendered in the well-nigh overwhelming problem of mere existence. This zeal for learning forced many a future colonial to include in his limited baggage a few books for the stimulation of his mind.[1]

The notion that colonial culture was bereft of scientific content has been dispelled by the scholarly efforts of a number of historians of American science.[2] The earliest immigrants had to content themselves with obtaining their science from such compendia as La Primaudaye's *The French Academy*.[3] However, as soon as the raw edges of frontier life became smoothed by more settled conditions, the colonists began to supplement their meager supply of books with more esoteric titles. As far as science was concerned, it was natural that astronomy should be of primary concern, not only to those interested in navigation and surveying, but also to the literate public, who needed a minimum amount of celestial information to interpret their almanacs and especially to understand the preaching of their spiritual and intellectual leaders. The Puritan clergy saw in the expanding orderly display of heavenly bodies stronger and stronger proof of God's wisdom,[4] and in

* University of San Francisco. This paper was presented before the University of California History of Science Dinner Club, Berkeley, California, 3 November 1964.

[1] Louis B. Wright, "The Purposeful Reading of Our Colonial Ancestors," *ELH, a Journal of English Literary History*, 1937, *4*: 85–91.

[2] Leonard Tucker, "President Thomas Clap of Yale College: Another 'Founding Father' of American Science," *Isis*, 1961, *52*: 55–56, gives an extended list of important contributions. See also, Samuel E. Morison, *The Intellectual Life of Colonial New England* (New York: New York University Press, 1956), pp. 241–275; Theodore Hornberger, "Puritanism and Science," *New England Quarterly*, 1937, *10*: 503–515; Perry Miller, *The New England Mind: From Colony to Province* (Cambridge, Mass.: Harvard University Press, 1953), pp. 437–446.

[3] Pierre de La Primaudaye, *The French Academy* (London, 1618). For the importance of this work, see Wright, *op. cit.*, pp. 91–92.

[4] Joseph Huntington (1735–1794) gives a succinct expression of this view in his *An Astronomical Diary; or an Almanack for . . . 1761*, by a Student at Yale College [Joseph Huntington] (New Haven, 1761), p. 15: "But to pro-

the appearance of unusual phenomena, such as comets,[5] the portents of divine wrath.[6] In these processes of education in the new astronomy, such great names as Copernicus, Tycho Brahe, Kepler, Galileo, and Newton became commonplace. But there was many a lesser figure who made his contribution to the dissemination of science in the fledgling outposts of civilization in the New World, and some of these have not received the full attention which they merit.

Pierre Gassendi (1592–1655) was a French priest of eclectic interests who wrote voluminously in literary, philosophical, scientific, and theological fields.[7] Although remembered today chiefly as a reviver of atomism,[8] in the seventeenth and eighteenth centuries his scientific reputation was due to his astronomical attainments. Descriptions and evaluations of his astronomy have appeared elsewhere,[9] and it shall suffice here to say that he was one of the best astronomers in France in the post-Galilean period. His observation of the transit of Mercury in 1631, his contributions in 1634 and 1636 to the production of the first lunar map, and his work on sunspots made him famous.[10] Although his contemporary and posthumous influences in Europe are well known, his name is mentioned only occasionally in studies of American science.

Gassendi's astronomy became known in America chiefly through his *Institutio astronomica . . . Cui accesserunt Galilei Galilei* Nuntius Sidereus, *et Johannis Kepleri* Dioptrice,[11] a textbook containing the lectures he gave while occupying the chair of astronomy at the Royal College in 1645. Its contents have been aptly described by Samuel E. Morison:

> Although Gassendi, for the sake of protecting his position in the church, made an ostensibly impartial exposition of the Ptolemaic, Tychonian, and Copernican systems, his preference for the last was so obvious that the *Institutio* provided the clearest and most convincing exposition of the new astronomy that any learned man had yet written; and the inclusion of two important works by Galileo and Kepler did much to spread their doctrines.[12]

ceed upon the Copernican and Newtonian scheme of astronomy . . . what wonders of wisdom, power, and grandeur open to our view!"

5 The origins of such beliefs are described by Andrew D. White, *History of the Warfare of Science with Theology in Christendom* (2 vols., New York, 1899), Vol. I, pp. 171–183.

6 As a typical example, see Increase Mather, *Heaven's Alarm to the World* (Boston, 1681).

7 Biographical and bibliographical information, evaluation and influence of his work, and references to earlier sources will be found in *Pierre Gassendi, 1592–1655, sa vie et son oeuvre,* Centre International de Synthèse (Paris: Albin Michel, 1955), articles by A. Adam, A. Koyré, G. Mongrédien, and B. Rochot. Further treatment is given in *Tricentenaire de Pierre Gassendi, 1655–1955,* Centre National de la Recherche Scientifique (Paris: Presses Universitaires de France, 1957).

8 R. B. Lindsay, "Pierre Gassendi and the Revival of Atomism in the Renaissance," *American Journal of Physics,* 1945, *13*: 235–242; G. B. Stones, "The Atomic View of Matter in the XVth, XVIth, and XVIIth Centuries," *Isis,* 1928, *10*: 460–462.

9 Pierre Humbert, "L'Oeuvre astronomique de Gassendi," *Actualités Scientifiques et Industrielles,* 1936, *6*, no. 378: 1–31; Seymour Chapin, "The Astronomical Activities of Nicolas Claude Fabri de Peiresc," *Isis,* 1957, *48*: 13–29; Lynn Thorndike, *History of Magic and Experimental Science* (8 vols., New York: Macmillan, 1923–1958), Vol. VII, Ch. 14.

10 Pierre Humbert, "Une lettre inédite de Gassendi," *Ciel et Terre,* 1940, *56*: 14.

11 There were at least six editions from 1647 to 1702, published in France, England, and Holland.

12 Samuel E. Morison, "The Harvard School of Astronomy in the Seventeenth Century,"

Among books of serious purpose, few are as ephemeral as textbooks. Consequently, it is to be expected that extant copies of the *Institutio* having autographed evidence of use in colonial and post-Revolutionary America are not numerous. Among the earliest which have come to light is the 1653 London edition in the library of the Boston Athenaeum, dated and inscribed with the names of three Harvard students. The signature of Thomas Shephard (1658–1685) appears as "Thomas Shephardus me jure tenet 9.12° 1675/6"; that of Ebenezer Brenton (1687–1766) is simply dated 1706; but Nathaniel Lindall (1708–1776) exhibited a propensity (common to some students in every age) of multiple inscription, for his name is written four times (two dates of 1726) on the title page and three times (two dates of 1727) on the verso of the last page of Kepler's *Dioptrice*. William Brattle (1662–1717), a Harvard graduate of 1680 and an effective and beloved tutor from 1686 to 1697, signed his name in 1705, indicating he owned the book after he had left the college for the ministry. It is significant that a span of over fifty years exists between the first and last student owner.[13] The United States Naval Observatory in Washington has a copy of the 1683 London edition with the signature "John Thomas His Book 1714." [14] The same edition is represented in the New York Society Library carrying the name of James Otis (1725–1783) on the title page. Otis, fiery figure of the Revolution, who received his B. A. from Harvard in 1743, was noted as an undergraduate for devotion to his books, even during vacation. He read widely in all fields in preparation for the M. A., which he received in 1746.[15] The Allegheny College Library received as a bequest from James Winthrop (1752–1821) [16] his large library, which included a copy of the 1683 edition with his signature on the title page. Winthrop (B. A., Harvard, 1769), son of the famous colonial astronomer John Winthrop (1714–1779), though possessing little of the scientific stature of his father, was interested in science and mathematics throughout his life. The signature of an earlier owner, Isaac Winslow (1709–1777),[17] appears on the flyleaf. After graduating from Harvard in 1727, he became a successful merchant and Loyalist. On the verso of the flyleaf is the inscription, "And: Eliot Liber Donum Isaaci Winslow." Eliot (1718–1778) obtained his Harvard degree in 1737, and was later one of the ornaments of public and theological life in Boston and heroic comforter of the sick and wounded in the siege

New England Quarterly, 1934, 7: 15. See also, B. Rochot, *op. cit.*, p. 41.

13 Information on Harvard graduates may be found in J. L. Sibley, *Biographical Sketches of Graduates of Harvard University, in Cambridge, Massachusetts* (3 vols., Cambridge, Mass.: C. W. Sever, 1873–1932), and continued by Clifford K. Shipton, *Sibley's Harvard Graduates* (8 vols., Cambridge/Boston: Massachusetts Historical Society, 1933–1960).

14 Thomas (1696–1737) graduated from Harvard in 1715, and a few years later settled into the life of a country gentleman in Marshfield, Mass. See Marcia A. Thomas, *Memorials of Marshfield* (Boston, 1854), p. 56.

15 William Tudor, *Life of James Otis* (Boston, 1823), pp. 7–9.

16 Brooke Hindle, *Pursuit of Science in Revolutionary America, 1735–1789* (Chapel Hill: University of North Carolina Press, 1956), p. 331; *Dictionary of American Biography* (New York: Charles Scribner's Sons, 1936), Vol. 20, p. 407.

17 Not to be confused with Isaac Winslow (1670–1738) of Plymouth. For the latter see *National Cyclopaedia of American Biography* (New York: James T. White & Co., 1898), Vol. I, p. 188. Thomas, *op. cit.*, 19–24.

of the city during the Revolution. James Logan (1674–1751),[18] one of the most outstanding colonial intellectuals, had a library rich in scientific holdings. Although most renowned for his botanical work, he was no mean astronomer, and he owned two copies of the *Institutio*, the 1653 London edition and the 1656 Hagae-Comitum edition. The latter has his signature on the flyleaf. Both books are in the Library Company of Philadelphia. Lastly we may mention that Adonijah Bidwell (1716–1784),[19] Yale graduate of 1740, minister, ship chaplain, and teacher, has his signature on the title page of a 1653 copy. The book is now in the Williams College Library.

Other lines of evidence showing the distribution of the *Institutio* will be mentioned briefly. On the premature death of George Alcock (1655–1676), Harvard medical student studying for his second degree, an inventory filed in the Suffolk County probate records lists an unspecified edition.[20] A list of duplicates obtained by Cotton Mather from the Harvard College Library in 1682 included a Gassendi.[21] John Usher, leading Boston bookseller who occupied a position in the trade "which made his orders excellent standards of measure," received from London in 1683 "2 Gassendus Astronimy."[22] Catalogues of Benjamin Franklin (1706–1790)[23] and William Byrd II (1674–1744) of Virginia[24] list "Gassendi Astronomica." Gassendi's *Opera Omnia* in six volumes were published in 1658 and again in 1727. The *Institutio* appears in Volume IV. This set was too expensive for widespread use in America, but a manuscript catalogue of the Harvard College Library lists the 1727 edition[25] and Thomas Jefferson (1743–1826) owned one published in 1658.[26]

It is of prime importance to realize that during the period being discussed here, "luxury" purchasing of books was economically impossible, except in the cases of a few wealthy merchants and planters.[27] For the majority, any buying of books without the purpose of reading them was extremely unlikely. Books were considered valuable possessions, to be passed on from one generation to the other, as attested to in wills and inventories, where books are itemized and appraised individually.[28] Consequently, we

[18] Frederick Brasch, "James Logan," *Proceedings of the American Philosophical Society*, 1942, *86*: 3–12; Hindle, *op. cit.*, 21–22.

[19] Franklin B. Dexter, *Biographical Sketches of Graduates of Yale College* (New York, 1885), pp. 639–640.

[20] Samuel E. Morison, "The Library of George Alcock, Medical Student," *Publication of the Colonial Society of Massachusetts*, 1933, *28*: 350–357.

[21] Clarence S. Brigham, "Harvard Library Duplicates, 1682," *Pub. Colonial Soc. Mass.*, 1915–1916, *18*: 407–417.

[22] Worthington C. Ford, *The Boston Book Market* (Boston: The Club of Odd Volumes, 1917), pp. 7, 130.

[23] Benjamin Franklin, *A Catalogue of Choice and Valuable Books* (Philadelphia, 1744).

[24] J. S. Bassett (ed.), *The Writings of "Colonel William Byrd, of Westover in Virginia, esq."* (New York: Doubleday, Page & Co., 1901), p. 441.

[25] *Supplement to the First Alphabetical Catalogue of 1765 [1768–1770]*, Harvard University archives.

[26] E. Millicent Sowerby, *Catalogue of the Library of Thomas Jefferson* (5 vols., Washington, D. C.: Library of Congress, 1959), Vol. V, p. 165.

[27] Howard M. Jones, *America and French Culture, 1750–1848* (Chapel Hill: University of North Carolina Press, 1927), p. 175.

[28] George K. Smart, "Private Libraries in Colonial Virginia," *American Literature*, 1938, *10*: 25; Thomas G. Wright, *Literary Culture in Early New England, 1620–1730* (New Haven: Yale University Press, 1920), *passim*; C. A. Herrick, "The Early New Englanders: What Did They Read?," *The Library*, 1918, *9*: 1–6.

may be assured that the astronomy of Copernicus, Kepler, and Galileo had ample assimilation into the thoughts of the owners of the *Institutio astronomica*; of course, it must be remembered that the extant copies represent only a fraction of those imported, for, inevitably, most of them disappeared.

Since Gassendi's book was a text, and a well-written one, it is not surprising to find it adopted in colonial colleges. While a student at Yale, Jonathan Edwards (1703–1758), later famous as the leader of revivalism in 1740–1741 and one of the most profound philosophical minds ever to appear in America, wrote a letter to his father on 21 July 1719, informing him of the books needed for the coming academic year, and among those listed was Gassendi's astronomy.[29] At Harvard, the earliest official mention of Gassendi's book is in the program of studies for 1723, where we learn that in the "fourth year the Senior Sophisters recite . . . Gassendus's astronomy."[30] Morison infers the probability that this program had been in effect since late in the seventeenth century.[31] However, there are two lines of inference by which the date of the first use of Gassendi may be pushed back still further. One has already been mentioned, namely that Shephard in 1675 and Alcock in 1676, while students, owned copies of the 1653 edition; and of course, this does not mean they were the first two. Another factor to consider is the strong influence on Harvard of English educational practice as carried on at Cambridge and in the dissenting academies.[32] Gassendi was used at Cambridge[33] and the academies for many years.[34] Norton notes the suggestive resemblance between some of these academies and Harvard in their curricula and texts.[35] Hence, considering the English background and the surviving evidence, it is quite probable that Harvard students were studying Gassendi at least as early as 1675, and that they continued to do so for at least half a century. As mentioned earlier, Shephard's book, passing to Brenton and Lindall, spans precisely this period, and while it is a unique surviving example, it seems to bear witness to the general practice.

In the introduction it is mentioned that practically everyone would have been interested in acquiring astronomical information to the extent of his capabilities for the purpose of interpreting the yearly almanacs as intelligently as possible in early colonial times. The almanacs, in addition to

[29] Thomas H. Johnson, "Jonathan Edward's Background of Reading," *Pub. Colonial Soc. Mass.*, 1930–1933, *28*: 197. It was precisely at this time that Rector Cutler (1684–1765) added the *Institutio* to help the students understand Newton and Locke (Edwin Oviatt, *The Beginnings of Yale (1701–1726)* [New Haven: Yale University Press, 1916], p. 423). In the Yale Library there is a one-page sheet (undated) compiled by J. C. Schwab, "A Partial List of the Text Books Used in Yale College in the Eighteenth Century," in which Gassendi is recommended for seniors in 1720. See Schwab's "The Yale Curriculum 1701–1901," *Educational Review*, June, 1901, p. 4.

[30] Arthur Norton, "Harvard Text-Books and Reference Books of the Seventeenth Century," *Pub. Colonial Soc. Mass.*, 1930–1933, *28*: 366.

[31] Samuel E. Morison, *Harvard College in the Seventeenth Century* (2 vols., Cambridge, Mass.: Harvard University Press, 1936), Vol. I, p. 146.

[32] Hindle, *op. cit.*, p. 86.

[33] Phyllis Allen, "Science in English Universities in the Seventeenth Century," *Journal of the History of Ideas*, 1949, *10*: 235.

[34] Herbert McLachlan, *English Education under the Test Acts; Being a History of Nonconformist Academies, 1662–1820* (Manchester: University Press, 1931).

[35] Norton, *op. cit.*, p. 376.

their utilitarian content, were the most widely circulated forms of general cultural information, for they almost always carried some kind of literary contribution, be it poetic, legal, historical, or scientific. Naturally, some of them contained brief discourses on the new astronomy. An example is the almanac of 1659 compiled by Zecharia Brigden (1639–1662),[36] a Harvard graduate (A. B., 1657) and tutor, who gives a brief exposition of the Copernican system, including four propositions which can be used to answer "The Objections usually brought to evert the truth of this Systeme." Each proposition is buttressed by such authorities as Kepler and Galileo, and others. The fourth proposition states "That a common motion of bodyes can imprint no force at all, to hinder, or further their different particular motions . . . *Gassend.*" Brigden then explains that the Scriptures are written for the common man as well as for the learned philosopher, and therefore "in Philosophicall truths therin contayned, the proper literal sense is always subservient to the casting vote of reason . . . *Gassend.*" Two years later another Harvard graduate, Samuel Cheever (1639–1724, B. A., 1659), concluded his almanac with an essay on the rise and progress of astronomy, in which the reader is informed that recently, "*Galilaeus, Bullialdus, Keplerus, Gassendus* and sundry other Mathematicians, have learnedly confuted the Ptolemaick and Tychonick System, and demonstrated the Copernican Hypothesis to be most consentaneous to truth and ocular observations, . . ."[37] Thus these early almanacs served as banners proclaiming the Copernican system. Emblazoned by the names of the popular Gassendi and the more illustrious Kepler and Galileo, they penetrated towns and remote farms throughout all New England. In this manner Gassendi assisted in the wide diffusion of the new ideas, and although the almanacs as sources of astronomical information were not perfect, at least (as expressed neatly by Morison)[38] they did "good work in persuading the New England farmer that he lived on a revolving planet."[39]

Two of the most influential Puritan divines and authors, Increase Mather (1639–1723) and his son Cotton (1663–1728), demonstrate their familiarity with Gassendi the astronomer. In his treatise on comets, Increase explains that the blaze or coma of the comet is caused by sunlight, as discerned by "*Tycho,* and of later times *Gassendus,* and many other famous mathematicians."[40] Here we have another example of a seventeenth-century American writer classifying Gassendi among the giants of astronomy. On 11 December 1719 a spectacular aurora was observed throughout New England. Cotton Mather thought it desirable to preserve its description for posterity and to make a few speculations on its nature, and therefore

[36] Zecharia Brigden, *New England Almanack* . . . (Cambridge, 1659), pp. 14–15.

[37] *An Almanack for the Year of Our Lord 1661,* by S. C. Philomathemat [Samuel Cheever] (Cambridge, 1661), p. 16.

[38] Morison, *Harvard College . . . , op. cit.,* Vol. 1, p. 217.

[39] There is no evidence that Brigden and Cheever were acquainted with Gassendi's *Institutio.* Morison (*Harvard College . . . ,* Vol. 1, p. 218) inclines to the belief that Harvard students at this time knew Gassendi only through Vincent Wing's *Astronomia instaurata* (1654) and Adrian Heerebord's *Meletemata philosophica* (1654).

[40] Increase Mather, *Kometographia* (Boston, 1683), pp. 14–15.

he quickly committed his observations to writing and had them published before the end of the year in a sixteen-page pamphlet. Gassendi had observed a similar celestial display in Aix, France, on 12 September 1621, and in characteristic fashion described it very fully and carefully, and, moreover, gave it the name "aurora borealis."[41] So well established was Gassendi's position in astronomy and so well known in America, that even a century later Cotton Mather in his brief treatise leans on the authority of Gassendi. Mather mentions the latter's name in three places, calling attention to the similarity between the observations in New England and those in Aix.[42] Since the Mathers were intellectual and spiritual leaders of stature, their writings were widely read, and through these media Gassendi's work in astronomy was disseminated.

Another clerical line of transmission of Gassendi's astronomical influence was provided in the person of Samuel Lee.[43] A fellow and later dean (1653) of Wadham College, Oxford (at that time the most active scientific center in England), he became infused with the new science, and was thoroughly familiar with Gassendi's writings. Lee wrote a number of books designed to illustrate the limitations of science and the use of its discoveries for the glorification of God, a common intellectual posture of the period. He migrated to Boston in 1686, and exerted a strong influence on the scientific thinking of leading colonials. Some of his books were reprinted in Boston. In the *Triumph of Mercy in the Chariot of Praise*[44] Lee displays his vast familiarity with the science of his time by quoting numerous past and contemporary authorities. When it comes to astronomy, he is overwhelmed by the immensity of the universe, and tries to convey to the reader some conception of the minuteness of the earth by giving various astronomical dimensions. For the latter he acknowledges that his figures came from Gassendi.[45] After Lee's death his library was offered for sale at public auction. The catalogue was the first of its kind to appear in America, and three of Gassendi's publications are listed, two of them being of astronomical interest.[46]

A few of Gassendi's contributions outside the field of astronomy were available to early Americans. Harvard students for about forty years beginning with 1687, in using Charles Morton's (1627–1698) *Compendium physicae,* were informed that "Gassendus and other Atomists will have but

41 Humbert, "L'Oeuvre astronomique de Gassendi," *op. cit.*, p. 28.

42 *A Voice from Heaven. An Account of a Late Uncommon Appearance in the Heavens. With Remarks upon It* (Boston, 1719), pp. 3, 4, 6. Publication was anonymous, but authorship is established in Thomas J. Holmes, *Cotton Mather, a Bibliography of His Works* (3 vols., Cambridge, Mass.: Harvard University Press, 1940), Vol. III, p. 1189.

43 Lee's important role in America is treated fully by Theodore Hornberger, "Samuel Lee (1625–1691), a Clerical Channel for the Flow of New Ideas to Seventeenth-Century New England," *Osiris*, 1936, *1*: 341–355.

44 Samuel Lee, *Triumph of Mercy in the Chariot of Praise* (London, 1677; reprinted Boston, 1718).

45 *Ibid.* (Boston reprint), p. 126.

46 *Library of the Late Reverend and Learned Mr. Samuel Lee* (Boston, 1693). *De Apparente magnitudine Solis humilis* is listed on p. 9, and *De Nicolai Claudii Fabrici de Peiresc . . . Vita* is on p. 13 (see sources in fn. 9 for importance of Peiresc in astronomy). Also listed on p. 13 is *De Vita et Moribus Epicuri libri octo.*

one kind of matter for all things (which they call corpuscles or little bodyes) which (say they) are variously shaped, and have between them little vacuities."[47] Although Morton exposes his students to this theory, he probably does so merely because he thinks that an educated gentleman should know that it is abroad. However, he himself was not an atomist, and did not believe in the existence of a vacuum, and consequently his noncommittal treatment probably did not convert any Harvard men to corpuscularianism. Samuel Lee, drawing an analogy between the spiritual force of God and the force which impels a bullet, makes a brief reference to Gassendi's *De motu impresso a motore translato* (1642).[48]

Even in the late eighteenth century and on into the nineteenth we have indications that Gassendi's reputation still lingered. Among those involved in the phlogiston controversy in America, Samuel L. Mitchill (1764–1831), professor of chemistry at Columbia College, at first tried to play the role of mediator between Priestley and his opponents. In a letter to Priestley dated 14 November 1797, and published in the *Medical Repository*, 1798, *1*: 514, Mitchill wrote:

> Perhaps even now my labours are but of little avail, or, if they were capable of bringing about a coalition of parties, I might say to you, after all, in the words of Prior in his Alma:
>
> For, Dick, if we could reconcile
> Old Aristotle with Gassendus
> How many would admire our toil!
> And yet how few would comprehend us![49]

Samuel Miller (1769–1850), noted Presbyterian clergyman and educator, in his highly regarded *Brief Retrospect of the Eighteenth Century*, saw fit to mention that the aurora borealis was named by the "celebrated Gassendi."[50] Nathaniel Bowditch (1773–1838), famous New England navigator, mathematician, and author of the still standard *New American Practical Navigator*,[51] taught himself French and Latin, and read widely in astronomy and mathematics. Even though interested primarily in the works of Lacroix and Laplace (he translated the latter's *Mécanique céleste* between

[47] Theodore Hornberger (ed.), "Morton's *Compendium physicae*," *Pub. Colonial Soc. Mass., Collections*, 1940, *33*: 36. Morton came from England to Harvard in 1686, became a minister at Charlestown, a fellow (1692) and vice-president (1697) of Harvard College.

[48] Samuel Lee, *Joy of Faith* (Boston, 1687), p. 231.

[49] Quoted in Sidney Edelstein, "The Chemical Revolution in America," *Chymia*, 1959, *5*: 164–165. Matthew Prior (1664–1721), English poet and diplomat, was an Epicurean, and hence it is not surprising that he was familiar with Gassendi. The poem, "Alma: or, the Progress of the Mind," may be found in *The Literary Works of Matthew Prior*, H. B. Wright and Monroe Spears (eds.) (2 vols., Oxford: Clarendon Press, 1959), Vol. I, p. 477. It had for its model the more famous satire, "Hudibras," by Samuel Butler (1612–1680).

[50] Samuel Miller, *Brief Retrospect of the Eighteenth Century* (New York, 1803; London, 1805). The book merited wide circulation. The quotation refers to the London edition, Vol. I, p. 253.

[51] Nathaniel Bowditch, *New American Practical Navigator* (Newburyport, Mass., 1802). This book has been and continues to be revised every few years. See Dirk Struik, *Yankee Science in the Making* (Boston: Little, Brown, 1948), pp. 72–77, 172–176, 390.

1814 and 1817), he owned a copy of the 1683 edition of Gassendi's *Institutio astronomica.* A further indication of Gassendi's reputation in America is to be found in the fact that he was selected for inclusion in *New Biographies of Illustrious Men.*[52] The contents of this book were chosen from the eighth edition of the *Encyclopaedia Brittanica.* Gassendi's biography occupies fifteen pages and was written by Henry Rogers (1806–1877), at one time professor of English language and literature in University College, London, and one of the ablest contributors to the *Edinburgh Review.*

It is evident that Gassendi's influence in America was widespread, being very active for at least fifty years following 1675, and not dwindling completely until the commencement of the nineteenth century. His ideas reached all classes — including farmers and people in remote towns — by means of early almanacs, college students who upon graduating entered various occupations, and a variety of intellectual leaders. Mainly through his *Institutio astronomica,* Gassendi assisted in the gradual erosion of Aristotelianism, and paved the way for the promulgation of Copernican and finally Newtonian science.[53]

52 Boston, 1857, pp. ix–xi, 289–304.

53 Louis B. Wright, *The Cultural Life of the American Colonies, 1607–1763* (New York: Harper, 1957), p. 220.

David Rittenhouse and the Illusion of Reversible Relief

By Brooke Hindle & Helen M. Hindle***

IN 1786, David Rittenhouse published an important study in experimental psychology, entitled "Explanation of an Optical Deception," which has been neglected or misunderstood by historians and psychologists alike.[1] Rittenhouse's scientific work has not been satisfactorily evaluated in any of its phases and this paper offers one opportunity to gauge his capacity and influence. It also suggests that there may have been more American work in experimental psychology before the late nineteenth century than is generally believed.

Rittenhouse's extensive reputation among his American contemporaries as a leading scientist is well known. It was attested by Benjamin Rush who called him "one of the luminaries of the eighteenth century" and by Thomas Jefferson who considered him the greatest living astronomer.[2] Perhaps his elevation to succeed Benjamin Franklin as president of the American Philosophical Society was the most significant honor he received. His reputation was based primarily upon his remarkable orrery, or mechanical planetarium, and upon his observations of the transit of Venus in 1769. His interests, however, were broad but generally related, astronomy and mathematics being at the focus of his attention. Both of them had some relationship to his calling as a clockmaker and mathematical instrument maker. In this connection, too, he ground lenses, constructed telescopes, and became interested in the laws of optics. Indeed, he thought at one time of offering a course of lectures in optics. He performed various experiments on the behavior of light, and certainly one of his most important contributions was his description of a diffraction grating in which he recorded the angular displacement of the six orders of spectra he could observe with it.[3]

In 1774, Rittenhouse began a series of experiments which he believed to be related to the behavior of light and the laws of optics. Only as he proceeded did it become clear that he was dealing with human behavior, or psychology, instead. The topic he investigated was an "optical deception" encountered in the use of microscopes and telescopes, later described as the

* New York University.

** Ho-Ho-Kus School System.

1 David Rittenhouse, "Explanation of an Optical Deception," American Philosophical Society, *Transactions,* 1786, *2:* 37-42.

2 Benjamin Rush, *Eulogium Intended to Perpetuate the Memory of David Rittenhouse* (Philadelphia, [1796]), p. 24; [Thomas Jefferson], *Notes on the State of Virginia* ([Paris], 1782 [1785]), p. 119.

3 David Rittenhouse, "An Optical Problem, Proposed by Mr. Hopkinson, and Solved by Mr. Rittenhouse," American Philosophical Society, *Transactions,* 1786, *2:* 201-206; Brooke Hindle, *The Pursuit of Science in Revolutionary America, 1735-1789* (Chapel Hill, 1956), pp. 169-71, p. 346.

cameo-intaglio illusion or the illusion of reversible relief. The effect of this illusion is to reverse the relief of a three-dimensional surface, to make the raised parts appear depressed and the depressed parts raised. His investigation of the illusion was conducted carefully by varying the conditions and recording the data in such a way that others could repeat the work. Even though he served as his own subject and obtained his data through introspection, he performed a genuine psychological experiment.

The spilling over of a problem in physics to the realm of psychology is not surprising in the field of vision. Investigators ever since the fifteenth century had been using optical instruments in their search for the laws of optics, and the human factor in the use of these instruments readily occurred to them. Even the analysis of the human eye as an optical instrument was undertaken early. The surprising thing about this particular psychological experiment is its inadequate evaluation by subsequent experimenters and the scant notice it has received from historians.

Rittenhouse's interest in reversible relief was *not* awakened by work then in progress in Europe. His stimulus, in so far as he had one beyond his natural curiosity in optics, was the paper submitted in 1745 to the Royal Society of London by Philip Frederick Gmelin of Wurtemberg and published in English in 1756. Gmelin's study, undertaken upon the suggestion of a friend, revealed the reversal of relief in seals, carvings, and other objects observed through both microscopes and telescopes. He varied the position of the instrument with respect to the object as well as the position of the object in relation to the light source, noting that he did not always see the depressed parts raised and the elevated parts sunk and that he did not always see the same conditions of relief seen by a friend. His experiments led him to "suspect, that all those fallacies were owing to shade," but he concluded, "why all these things happen exactly after the same manner, I do not pretend to determine."[4]

The methodical experimenters of the nineteenth century found numerous records of other related activity before Rittenhouse began his work. From the earliest times, artists were concerned with the problem of relief, but *reversal* of relief seems first to have been observed and recorded at a meeting of the Royal Society in 1668. On that occasion some of the members in examining a coin with a new microscope saw the image depressed but others perceived it as raised. This incident was recorded in Thomas Birch's *History of the Royal Society* where Rittenhouse may have come upon it.[5]

In 1718, Louis Joblot, Professeur Royal en Mathématiques de l'Académie Royale de Peinture et Sculpture, confronted the problem of reversible relief in a work on microscopes and telescopes. All his observations depended, he said, on the fact that in these instruments "objects . . . are not seen directly but by reflected light."[6]

Much less pertinent, though occasionally cited by later researchers, was the work of Bonaventure Abat who in 1763 had written on an optical illusion

[4] P. F. Gmelin, "Of Fallacious Vision through Compound Microscopes," John Martyn, ed., *Philosophical Transactions Abridged*, 1756, *10:* 31, 32. (Originally appeared in Latin in *Phil. Trans.*, 1745, *43:* 385-391.)

[5] Thomas Birch, *The History of the Royal Society* (London, 1756-57), II, pp. 348-49.

[6] Louis Joblot, *Descriptions et Usages de Plusiers Nouveaux Microscopes* (Paris, 1718), p. 94. This quotation and all others not published in English have been translated by the authors.

caused by the inversion of the image. This was not a question of reversible relief, however, but a very particularized illusion in which the image of a half-filled bottle when inverted by a concave mirror still appeared to have water at the bottom. The illusion he believed to be a natural judgment independent of will or reason resulting from the subject's "never seeing water suspended over air in a flask but always at the bottom."[7] This conclusion was a psychological explanation emphasizing the role of past experience in perception.

It is unlikely that Rittenhouse knew anything of the work of Joblot or Abat, but even if he had, he would not have found the answers he sought. "The cause of this appearance," the American exclaimed, "for any thing I know, remains still to be explained."[8] And indeed, this was so.

Rittenhouse began his experiments by placing two convex lenses in a tube at, or close to, the sum of their focal lengths and observing a series of three-dimensional objects. Always the image was inverted and often the raised portions appeared depressed and the depressed portions raised. Thinking about it, he discarded the possibility that the reversed relief depended only upon the laws of optics, although his initial notion was that reversal was caused by the near points of the object becoming the far points in the optical image. He was next led to the thought that the reversal must be a result of the *interpretation* of the inverted image, and this idea was strengthened when he considered the appearance of the moon through a telescope. To his trained eye, cognizant of the source of light, the heights and valleys of the surface were well defined. To the untrained observer, nothing but dark and light patches appeared.

These thoughts led to the hypothesis that the illusion depended upon rules of judging "imperceptibly formed in the mind."[9] He postulated that the key to the perception of the relief of a three-dimensional surface was the interpretation of shadows in relation to the perceived direction of illumination. To test this, he looked through his inverting tube at a brick hearth illuminated from a single window opposite the chimney. As expected, the relief reversed, "all the bricks appeared depressed and the clefts between them elevated."[10] Then when he reversed the actual direction of incident light on the hearth with a mirror placed against the chimney and a board to screen the direct light from the window, the relief reversed again to appear in its normal condition with the mortar depressed and the bricks raised.

A second experiment revealed that a tube without lenses produced the reversed relief illusion when the direction of incident light was altered 180° by the mirror. The surface assumed its normal appearance when the light actually came from the window. These experiments supported his hypothesis that the illusion was a result of miscuing the interpretive faculties of the mind by a perceived direction of light different from the actual direction.

Rittenhouse went on to investigate the strength of the illusion, discovering that the insertion of a finger, pen, or pencil in the field was enough to break the illusion and cause the surface to be interpreted correctly. After breaking the illusion in this manner once or twice, he found it difficult to reestablish

[7] Bonaventure Abat, *Amusmens Philosophiques sur Diverses Parties des Sciences* (Amsterdam, 1763), p. 246.

[8] Rittenhouse, "Explanation of an Optical Deception," p. 37.

[9] *Ib.*, p. 38.

[10] *Ib.*, p. 39.

even though it would always return if it had been destroyed by removing the lenses.

He had attacked a limited problem in human perception experimentally and had found the answers to his specific questions—this a century before the founding of the first psychological laboratory. It is true that later experimenters discovered many cues for interpreting the character of a three-dimensional surface in addition to the pattern of light and shade—perspective, convergence, and retinal disparity to mention a few. Furthermore, since the latter part of the nineteenth century, psychologists have investigated a great variety of optical illusions and have accounted for them by a multitude of theories. Some theories emphasize the stimulus, others the individual; some stress the physiological factors, others the psychological; some assert the importance of relationships within the stimulus, others the role of past experience. Rittenhouse stood at the beginning of this development. His experiment placed the illusion of reversible relief firmly in the larger field of perception and underlined the contribution of past experience in this process. His conclusions have had permanent validity.

The lack of influence of his work is perhaps the most striking thing about it. There is no indication that the publication of his research in 1786 in the *Transactions* of the American Philosophical Society had any impact upon anyone. The *Transactions* were pretty well distributed throughout the United States and known in some of the European centers, but as no one had been interested in the problem when Rittenhouse took it up, no one was interested when he solved it. Rittenhouse's work served to stimulate further research no more than Gmelin's had. When he died, ten years later, the only published reaction to this successful work was to be found in a few unenthusiastic reviews.[11]

Not until forty years after Rittenhouse's publication did a scientist pick up his studies. David Brewster at the University of Edinburgh was led in 1826 to extend his studies of optics to the consideration of just the problem that had engaged the American. There is no evidence that his inquiry was stimulated by Rittenhouse's paper, but he certainly made extended use of it and even admitted that he found there "a correct explanation of the illusion."[12] Indeed, Brewster's first paper on the subject was little more than a recapitulation of Rittenhouse's work—without complete acknowledgment of the extent to which he borrowed. His experiments differed in form but not in substance from those performed by Rittenhouse long before.

A few years later in Germany, Georg Muncke criticized the work of both Rittenhouse and Brewster in a general article on vision. He reported that Rittenhouse had explained the illusion as a result of "inverted incident light," an explanation he could not accept.[13] Here Muncke did less than justice to Rittenhouse by representing a part of his explanation as the whole of it. He asserted that the inverting lenses used by Rittenhouse and Brewster obscured the problem while his simple magnifying glass isolated it. He failed to

[11] *Critical Review,* 1787, *64:* 242, 444-45; *Monthly Review,* 1786, *76:* 143.

[12] David Brewster, "On the Optical Illusion of the Conversion of Cameos into Intaglios and of Intaglios into Cameos, with an Account of other Analogous Phenomena," *Edinburgh Journal of Science,* 1826, *4:* 103.

[13] Georg Wilhelm Muncke, "Gesicht," in Johann S. T. Gehlers, *Physicalische Wörterbuch* (Leipzig, 1828), p. 1455.

acknowledge that the American had produced the illusion, without inversion, by using an empty tube. Muncke discarded the conclusions of Rittenhouse and Brewster without seeming to understand their derivation and without offering an alternative solution. He admitted that in his terms, "the explanation yet stands incomplete." [14]

Subsequent criticisms of Rittenhouse's work were, like Muncke's, based either on inaccurate comprehension of his experiments or on the distinctly limited nature of the problem he investigated. Charles Wheatstone, British physicist, was unable to consider the illusion of reversible relief except in reference to the retinal disparity of binocular vision, the significance of which he established through experiments with the stereoscope. Wheatstone considered monocular vision, which alone had engaged Rittenhouse, as merely a special case of the larger problem of binocular vision. To Wheatstone, the illusion occurred because in monocular vision the cues available in binocular vision were lacking. The problem he studied was distinctly different from that reported on by Rittenhouse.[15]

Working at about the same time as Wheatstone, Hugo Schröder developed additional aspects of the general problem. He studied reversible perspective using his famous drawing of a staircase and reversible relief encountered with the unaided eye. He considered Rittenhouse's explanation insufficient—as it certainly was in terms of Schröder's broader problem which included the illusion of reversal in line drawings where shadows could not play a part. Yet, when he asserted that Rittenhouse's conclusions depended upon "an inversion of shadows through an optical apparatus," his digest, too, was incomplete and unfair. Schröder did admit, however, that the use of an optical tube by the subject was "one of a whole variety of conditions" that would alter any conclusions about reversible relief.[16] Moreover, he confirmed the significance of the factor of experience in perception, emphasized by Rittenhouse, when he investigated the reversal of relief in molds of objects normally perceived only as convex. Schröder found that the reversal from concave to convex came very easily with molds of human heads and animals but that the illusion was much more difficult to experience with meaningless three-dimensional decorations.[17]

In his monumental *Optics,* Herman von Helmholtz compiled a very revealing historical summary of the problem of reversible relief—revealing not so much in its facts as in its appraisal of the relative significance of Rittenhouse's work. Helmholtz reported, "As was noticed by Rittenhouse the illusion may be heightened and facilitated also by reversing the illumination of the matrix." [18] Then, he discussed the reversal of illumination in terms of the anaglyptoscope, an instrument invented in 1854 by Johan Jacob Oppel for reversing the incident light without the subject's knowledge.[19] It altered light

[14] *Ib.,* p. 1456.

[15] Charles Wheatstone, "Contributions to the Physiology of Vision, Part 1," *Phil. Trans.,* 1838, *128:* 371-94.

[16] H. Schröder, "Ueber eine optische Inversion bei Betrachtung verkehrter, durch optische Vorrichtung entworfener, physischer Bilder," in J. C. Poggendorf, *Annalen der Physik und Chemie,* 1858, *105:* 298, 299.

[17] H. Schröder, "Ueber eine optische Inversion mit freiem Auge," in J. C. Poggendorf, *Annalen der Physik und Chemie,* 1852, *87:* 308.

[18] Herman von Helmholtz, *Treatise on Physiological Optics,* 3d edn., trans. by J. P. C. Southall (Rochester, 1925), III, p. 287.

[19] J. J. Oppel, "Ueber ein Anaglyptoskop," in J. C. Poggendorf, *Annalen der Physik und Chemie,* 1856, *99:* 466-469.

direction by means of a screen and a mirror placed opposite the light source. This, of course, is exactly what Rittenhouse had done when he cut off the light from his window with a board and viewed the hearth in the light reflected from a mirror on the chimney! Yet Helmholtz made no note of Rittenhouse's initial "anaglyptoscope," nor had Oppel in his report on the apparatus. There is no indication that Oppel was acquainted with Rittenhouse's work, although most of his contemporaries working in the field were. However that may be, Helmholtz mentioned Rittenhouse but failed to acknowledge his original procedure for reversing illumination.

Helmholtz also followed Muncke in forgetting an essential part of Rittenhouse's study when he reported that "Rittenhouse tried to explain it as being due to a reversal of illumination."[20] He quoted Muncke's objection to Rittenhouse's conclusion, the fact that the illusion can also be seen in a simple magnifying glass, and then presented as a confirmation of Muncke's view, Abat's experiment with the half full bottle and the concave mirror. This was an experiment conducted before Rittenhouse and one which even Muncke considered in a category different from reversible relief.

Whether Helmholtz ever read Rittenhouse's paper may be seriously questioned but his evaluation has had extensive influence and has not provoked much dissent. Probably his emphasis on Oppel's work caused Charles N. Winslow as late as 1933 to introduce his experiments concerning visual illusions in chickens with the statement that "Oppel in 1854 is reported to have conducted the earliest experimental study of an illusion."[21] At any rate, Edwin G. Boring, in his influential study of the experimental psychology of sensation and perception first published in 1942, reduced Rittenhouse's contribution to a single sentence: "Rittenhouse noted in 1786 that a low relief or matrix may reverse its depth and appear like a raised relief or patrix, the concave turning convex."[22] This observation, of course, had been made long before Rittenhouse. More unfortunately still, Boring then assigned to Oppel precisely the most important discovery made by David Rittenhouse: "Oppel (1855) with an ingenious instrument he called an anaglyptoscope, showed that the perception depends, not upon the real direction of the light, but upon its perceived direction."[23]

Rittenhouse's study remains, within its own terms, a well-designed series of experiments which yielded new information and lead to new understanding. It solved the problem undertaken. Many factors account for the continuing failure of later investigators to evaluate his work properly. For one thing, it was isolated in time from all related research and can only be regarded as a forerunner of the active investigations of the nineteenth century. More important, it did not provide the answers to the questions of later experimenters —only to his own.

[20] Helmholtz, p. 362.

[21] Charles N. Winslow, "Visual Illusions in the Chick," *Archives of Psychology,* 1933, *153:* 5.

[22] Edwin G. Boring, *Sensation and Perception in the History of Experimental Psychology* (New York, 1942), p. 266.

[23] *Ib.,* p. 266.

An Attempt in the United States to Resolve the Differences between the Oxygen and the Phlogiston Theories †

By Robert Siegfried *

BEFORE the arrival of Joseph Priestley in America in 1794, there had not been any apparent concern among American chemists as to the phlogiston controversy that had been taking place in Europe. Priestley had become, by that time, the last major chemical figure still maintaining an undiminished preference for the phlogiston theory. But he neither found nor made any adherents

† Read before the History of Science Society, Baltimore, Maryland, 3 April 1954. This paper is based upon a portion of a Ph.D. thesis submitted to the graduate school of the University of Wisconsin. The author gratefully acknowledges the help and encouragement provided by Professors Aaron J. Ihde and Robert C. Stauffer of the University of Wisconsin, under whose direction this work was originally undertaken.

* Department of Chemistry, University of Arkansas.

to his views after his arrival.[1] The general lack of sympathy in the United States for Priestley's views on phlogiston, is indicative of the effectiveness of the introduction of the new chemistry before his arrival.[2]

However, when Priestley began publishing papers in the United States defending the doctrine of phlogiston, he succeeded in raising two competent opponents, John Maclean of Princeton and James Woodhouse of the University of Pennsylvania Medical School. Since accounts of the controversy between Priestley and these two men have appeared elsewhere,[3] I will not repeat the story here. While Maclean and Woodhouse were defending the anti-phlogistic viewpoint and attempting to refute the arguments of Priestley, Samuel L. Mitchill was making a serious attempt to promote a compromise position. Mitchill was a medical graduate of the University of Edinburgh in 1786, who taught at Columbia College (1792–1801) and served several years as United States Senator and Member of the House of Representatives. From the beginning of his teaching career, Mitchill taught the chemical doctrines of Lavoisier, and in 1794 he produced the first publication by an American on Lavoisier's chemical nomenclature. He thus played an early and influential role in the introduction of the antiphlogistic chemistry into America.[4]

Beginning in 1797, however, he proposed certain changes in the French nomenclature which suggest that he was not in full accord with the antiphlogistic doctrines. One of these changes was the substitution of the word phlogiston for hydrogen, by means of which he hoped to "accommodate the disputes among the chemists." It is with the significance of this change in terminology that this paper is largely concerned.

In 1797 Mitchill helped found the New York *Medical Repository*, the first medical journal in the United States, and served as the principal editor during the lifetime of the journal. As editor of this journal, Mitchill was in an unusually good position to attempt an arbitration of the phlogiston dispute just then starting.

The founding of the *Medical Repository* nearly coincided with the publication of Priestley's first American writings on phlogiston, and Mitchill immediately took advantage of the opportunity to publish the resulting discussion in his journal. In the first volume, there appear reviews of Priestley's "Considera-

[1] Priestley himself said, "In this country I have not heard of a single advocate for phlogiston." Quoted by Tenney L. Davis, Priestley's Last Defense of Phlogiston, *Journal of Chemical Education*, 1927, *4*: 178, from the preface to *The Doctrine of Phlogiston Established*, Philadelphia, 1800.

[2] For the details of the introduction of the chemical ideas of Lavoisier into the United States, see Denis I. Duveen and Herbert S. Klickstein, The Introduction of Lavoisier's Chemical Nomenclature into America, *Isis*, 1954, *45*: 278–292, 368–382.

[3] No single source seems to give the complete story. Edgar Fahs Smith's *Chemistry in America* (New York, 1914) is mostly concerned with the contributions of Woodhouse. William Foster's edition of Maclean's *Two Lectures on Combustion* (Princeton University Press, 1939) is understandably confined to Maclean's viewpoint. The best accounts of Priestley's efforts can be found in two papers by Tenney L. Davis: The Last Stand of Phlogiston, Priestley's Defense of the Doctrine after his Removal to America, *Studien für Geschichte der Chemie*, Festgabe für Edmund O. v. Lippman (Verlag von Julius Springer, 1927), pp. 132–147; and Priestley's Last Defense of Phlogiston, *Journal of Chemical Education*, 1927, *4*: 176–183.

[4] Duveen and Klickstein, *Isis*, 1954, *45*: pp. 281–289. The present paper was submitted to *Isis* just before the appearance of the paper by Duveen and Klickstein. I am indebted to the editor of *Isis* for the opportunity of slightly revising this paper by taking into account the information on the early efforts of Samuel L. Mitchill provided by Duveen and Klickstein.

tions on the Doctrine of Phlogiston," [5] Adet's answer to Priestley,[6] and Maclean's *Two Lectures on Combustion.*[7] Mitchill was also able to obtain the cooperation of Priestley in publishing his scientific writings in the *Medical Repository.* Priestley wrote to Mitchill on 14 June 1798,

> I am glad that your *Medical Repository* has been extended to subjects of general philosophy and chemistry. Had I known this before, I should have taken the liberty to send you an account of some of my late experiments, especially those which have for their object, the decision of the question between the Phlogistians and the Antiphlogistians.[8]

Priestley subsequently made the *Medical Repository* the principal outlet for his writings on phlogiston; these mostly took the form of letters to Mitchill which the latter then published. Altogether there are nineteen separate items by Priestley in the pages of the *Medical Repository* and ten more in the *Transactions* of the American Philosophical Society.[9] Nearly all these items are concerned with some aspect of the phlogiston dispute. Mitchill thus succeeded in localizing the discussion to the pages of his journal.[10]

Besides making the *Medical Repository* available for the writings of Priestley and his antagonists, Mitchill made a personal effort to reconcile the two opposing views. In the first volume, he printed a letter from himself to Priestley under the heading, "An Attempt to Accommodate the Disputes Among the Chemists Concerning Phlogiston." [11] In this letter, Mitchill took the position that, though oxygen is necessary for combustion, inflammable substances contain a principle of inflammability characteristic of all those substances which burn with a flame. Mitchill had noticed that a jet of steam when turned onto a piece of glowing charcoal caused the charcoal to glow with intensified color. Mitchill concluded that,

> . . . water underwent a true combustion and was inflammable, for the same reason that oil was, because it contained a something that would burn; and this something seemed to be exactly similar to that which made oil capable of exhibiting flame.[12]

After mentioning a few better examples of combustion, he stated,

> The circumstance common to all the processes I have mentioned, is 'burning with flame or blaze,' which, wherever it occurs, seems to indicate the presence of what has been called hydrogene. According to my present conception of the matter, this principle or substance, common to so many bodies and enabling them to

[5] *Medical Repository*, 1797, *1*: 221–225, 541–542.

[6] *Ibid.*, *1*: 225–229. P. A. Adet was at that time French minister to the United States. He translated Priestley's pamphlet and added his own critical answer to it, publishing them together in Philadelphia in 1797: *Réflexions sur la doctrine du phlogistique et la décomposition de l'eau.* Par Joseph Priestley. Ouvrage traduit de l'anglais et suivi d'une réponse par P. A. Adet.

[7] *Ibid.*, *1*: 348–350. These two lectures were printed in pamphlet form by Maclean in 1797: Two Lectures on Combustion supplementary to a Course of Lectures read at Nassau-Hall, containing an examination of Dr. Priestley's Considerations on the Doctrine of Phlogiston and the Decomposition of Water. These lectures, along with Priestley's Considerations, were reprinted under the editorship of William Foster in 1929, Princeton University Press.

[8] *Medical Repository*, 1799, *2*: 48.

[9] A complete list of Priestley's chemical writings published in America can be found in Tenney L. Davis, Priestley's Last Defense of Phlogiston, *Journal of Chemical Education*, 1927, *4*: 176–183.

[10] Priestley acknowledged the *Medical Repository* to be ". . . the theatre of the contest." Letter to Mitchill dated 30 January 1800, *Medical Repository*, 1800, *3*: 305.

[11] *Medical Repository*, 1798, *1*: 514–521.

[12] *Ibid.*, *1*: 514.

undergo inflammation, may in strict propriety, be called phlogiston. I always thought phlogiston a well-conceived word, and have objected to it not on account of the impropriety of the term as such, but because of the vague and unsatisfactory way in which it was used. If a definite signification can be affixed to it, I think the adoption of it would be still a great acquisition to philosophical language, and have a tendency to settle at least half the controversy which divides the chemists.

I propose, then, to expunge hydrogene and substitute *phlogiston* in its place. *Phlogiston* will thus be the radical term, and mean strictly the thing, in combustible bodies, which forms blaze or ignited vapour. The union of this with caloric, will make *phlogistous* or inflammable air, the air which burns with a blaze. The combination of phlogiston with oxygene, will constitute water or the *oxyd of phlogiston*, one of the products of inflammation, and like fixed air and other compounds, formed during the same process, incombustible in common temperatures and circumstances, afterwards.[13]

In further illustration of his views, he said, "Common brimstone . . . is not a simple substance, but is a *phlogisture* of sulphur." [14]

He held a similar view of phosphorus, but not necessarily of coal or charcoal, for they do not always burn with flame. He developed these ideas more fully a few years later in a pamphlet entitled, "Explanation of the synopsis of chemical nomenclature and arrangements containing several important alterations of the plan originally reported by the French academicians." [15] In this pamphlet he stated that the term phlogiston

. . . means all those atoms in bodies which *burn with flame or blaze.* This quality distinguishes them from atoms of carbone, which, if pure, burn indeed, but without any blaze whatever. It is the basis of fire-damp or inflammable air, and enters largely into the ordinary composition of sulphur, phosphorus, and metals, giving to them the power of burning *with flame.* Phlogiston is a plentiful ingredient in animal and vegetable bodies, and evidently enters into the composition of water; these being all capable of exhibiting blaze as they burn. From its being a constituent part of water, it has been called hydrogen, or the water-getter, and is distinguished by that name in all the modern books. But as generic names ought to be taken from the most obvious quality which any assemblage of atoms possesses, and as the exhibition of blaze is a more glaring appearance than the formation of water, the former deserves the preference in giving a title to the genus.[16]

Since phlogiston was to him contained in all inflammables, such things as sulphur and phosphorus were composed of the phlogiston plus some as yet undiscovered basis.

. . . phosphorus . . . as prepared in the laboratories . . . seems to be connected with much phlogiston, enabling it to burn with blaze, and form water in the process, after the manner of brimstone. The term, then, like all the rest, applies to phosphoric atoms, disengaged from all combination whatsoever; but as they were never known to exist in such a state of separation, the name refers to that state of abstract purity in which the mind may conceive to be.[17]

Thus we see that though his system is based upon the recognition of the role of oxygen in combustion, he felt the need for a principle of inflammability.

[13] *Ibid.*, 1: 515.

[14] *Ibid.*, 1: 517.

[15] New York, 1801, 44 pp. There are many ideas in this pamphlet besides those dealing with phlogiston, but they are not relevant here.

[16] *Loc. cit.*, p. 32.

[17] *Ibid.*, p. 36.

This principle he approximately identified with hydrogen, as had Cavendish and Kirwan before him. This meant to Mitchill that all things which burned with flame must contain hydrogen and that all things which contained hydrogen must burn with flame. In his concern for logical consistency, he continued to list water among the inflammable bodies.

Mitchill's first letter to Priestley embodying his attempted compromise, was answered by Priestley in a letter printed immediately after Mitchill's own.

> I thank you for your ingenious, and well intended attempt to promote a peace between the present belligerent powers of chemistry; but I fear your labor will be in vain. In my opinion there can be no compromise of the two systems.[18]

Thus Mitchill's scheme was rejected by Priestley, and everyone else apparently ignored it. After the publication of his pamphlet on chemical nomenclature in 1801, Mitchill wrote no more on the subject until 1810.

Supporting evidence for his system was found in the fact that many inflammables were known to contain hydrogen. The two most obvious exceptions were sulphur and phosphorus, and evidence was soon to appear that even these two contained hydrogen.

Humphry Davy, in the third Bakerian Lecture to the Royal Society, reported that sulphur when subjected to the heat of an electric arc yielded some hydrogen sulphide. Davy could account for the presence of hydrogen in no other way than to consider it a component part of the sulphur.

> The existence of hydrogen in sulphur is fully proved, and we have no right to consider a substance, which can be produced from it in such large quantities, merely an accidental ingredient.[19]

Davy reached analogous conclusions for phosphorus.

Dȧvy later withdrew the idea that hydrogen had been proved to be a constituent part of the sulphur and the phosphorus, but in his *Elements of Chemical Philosophy* published in 1812 he still considered the experimental data to be valid.

> Sulphur has been placed amongst the undecompounded bodies, because as yet nothing certain is known respecting its elements. When Sicilian sulphur was fused and exposed to the action of platina points intensely ignited by Voltaic electricity, excited by 1000 double plates, permanent gas was given off from it, which proved to be sulphuretted hydrogene: a small quantity of sulphuretted hydrogene is given off likewise during the action of copper filings and sulphur. . . . It may however be questioned whether hydrogene is essential to the constitution of sulphur. Surphur may possibly contain in its common forms a little moisture, or a little of a solid compound of hydrogene and sulphur; and till the gas can be separated from it in definite proportions, and be proved to be combined with some other matter, no accurate conclusion can be formed upon the subject.[20]

The first notice that Samuel L. Mitchill took of Davy's work of 1808 is found in the Medical and Philosophical Notes of Volume 13 of the *Medical Repository* (1810) under the heading, "Doctrine of Phlogiston Reviving." [21] On these three pages, he first gives a brief account of the history of the phlogistic and antiphlogistic theories. Without referring to himself by name,

[18] *Medical Repository*, 1798, *1*: 521.

[19] *Philosophical Transactions*, 1809, *99*: 62.

[20] *Loc. cit.*, pp. 283–284.

[21] *Loc. cit.*, pp. 289–292.

Mitchill describes his own early efforts in the following rather petulant phrases.

> In 1797, America offered to mediate between the philosophical belligerents; but Priestley was proud of contending alone against the whole host of anti-phlogistans, and the antiphlogistans were confident of numbers. . . . It was proposed to them to consider the hydrogen of the neologists as phlogiston, and to make the language conform by obliterating the former of these words, and substituting the latter in its place. But neither party would agree to this fair proposal. The war continued until it seemed in great measure to have ceased by the removal of the chief of the phlogistans from the fields.[22]

Two other notes of the same type appeared in the next volume. The first of these was entitled, "Further Symptoms of the Revival of Phlogiston," and contained a discussion of the dispute between Gay-Lussac and Davy. Gay-Lussac held that the metals potassium and sodium were compounds of hydrogen and the respective alkalis, while Davy contended that the alkalis were metallic oxides. Mitchill expressed himself clearly.

> It is probable the dispute admits of an easy accommodation, by merely considering the alkalis, &c. in their ordinary states, to be oxyds, or combinations of metallic bases with oxygen; and by viewing them in their reduced states to be combinations of metallic bases with phlogiston (hydrogen). This will keep up the analogy perfectly with the other metals. It will support the doctrine of Lavoisier, as far as oxygen is concerned in changing the forms and qualities of metals. It will do more: it will confirm the doctrine of Stahl, as far as phlogiston (hydrogen) is operative in changing and modifying their bases. It will happily reconcile the two systems, by allowing the facts that a metallic basis, by combining with oxygen, forms *an oxyd*, and by combining with phlogiston, forms what is called a *metal proper*. When deprived of both oxygen and phlogiston, that intermediate condition of the metal is produced, which most nearly approaches to the pure, uncombined, and simple element. These explanations may be seen on the chart of amended chemical nomenclature, published at New York, by Messrs. Swords, in 1801.[23]

In a later notice in the same volume of the *Medical Repository*, "Reconcilement of the Old and New Chemistry," Mitchill apparently felt that the experimental basis for his compromise stand was now adequately established, for he here presented his views more defiantly than before. In referring again to Davy's work on the presence of hydrogen in sulphur and phosphorus, he said,

> It seems at last to be admitted that sulphur contains hydrogen. This is so plain and palpable that nothing but the most determined opposition to the phlogistic theory would ever have ventured to deny it.[24]

Mitchill was not alone in America in trying to effect an alternative to the completely anti-phlogistic schemes. A major effort toward compromise was made by John Redman Coxe, the successor to James Woodhouse in the University of Pennsylvania Medical School. This effort was embodied in a pamphlet entitled, "Observations on combustion and acidification with a new theory of these processes, founded on the conjunction of the phlogistic and anti-phlogistic doctrines." [25]

[22] *Ibid.*, pp. 290–291.

[23] *Medical Repository*, 1811, *14*: 84–85.

[24] *Ibid.*, 1811, *14*: 186.

[25] Philadelphia, 1811, 50 pp. In order to read

Coxe's proposals though different in some details from those of Mitchill, are based largely on the same experimental "facts" of Davy. Coxe formalized his indebtedness to Davy by dedicating the pamphlet to him. This dedication reads in part:

> The splendid offerings you have so largely made to Science by your late important discoveries, sufficiently point out the propriety of dedicating to you, the following pages, which are founded chiefly on the results of your own experiments.

Coxe's scheme was similar to Mitchill's in that he recognized the role of oxygen in the combustion process, but also insisted upon a principle of inflammability, which he too identified with hydrogen.

Mitchill, perhaps encouraged by the company of Coxe in this stand, took the opportunity to say:

> As to the existence of phlogiston as the material of flame, it is such a self evident fact, that it is truly surprising that any persons possessed of their senses could have doubted it. And the employment of a magnificent galvanic apparatus, by a most skillful hand, to show that it exists in sulphur and phosphorus, is like proving from elaborate and optical considerations that men can see: or by establishing by grave and profound examinations of their component elements that stones are hard.[26]

This is almost a position of no retreat, and it seems inevitable that he must have regretted assuming it. His only subsequent statement in the *Medical Repository* appeared in 1813, and in it, perhaps chastened by the notable lack of enthusiasm for his views from other chemists, he stated,

> Much indeed must be done, in explaining the phenomena of combustion, and especially the blaze so frequent in burning bodies, before we can deny, or even doubt the existence of a phlogistic principle, common to that part of processes by fire, in which flame is conspicuous.[27]

This is a distinctly more judicious expression of his views than that quoted earlier, though he obviously had not changed his basic position.

Since both Mitchill and Coxe based their revival of a phlogistic philosophy of chemistry largely on the work of Humphry Davy, it is worth while to examine Davy's views more fully. This period of Davy's career has been described by one of his biographers as "a period of perplexities." [28] Davy's experimental work between 1808 and 1810 was indeed perplexing, but his interpretations were a good deal more cautious than those of Samuel L. Mitchill. Davy recognized the possibility of a phlogistic interpretation of his work, but nonetheless refrained from considering the system adequately demonstrated. On the other hand, he did not at that time actually reject it. In his *Elements of Chemical Philosophy* published in 1812, he developed certain ideas of chemical combination based upon the presumed presence of hydrogen in combustible bodies.

---

the copy contained in the University of Wisconsin Library, I had the unexpected task of slitting the pages. This copy, at least, made no converts to Coxe's views.

[26] *Medical Repository*, 1811, *14*: 366.

[27] *Ibid.*, 1813, *16*: 373.

[28] Joshua C. Gregory, *The Scientific Achievements of Sir Humphry Davy* (London: Oxford University Press, 1930), chapter 4, pp. 58–72, which bears this phrase as its title, contains the best account and interpretation available of this part of Davy's work.

> I have already hinted at the idea that all inflammable matters may be similarly constituted and may contain hydrogene, and on this supposition they may be conceived to owe their powers of combining both with oxygen and chlorine, to the attractive energies of their combined hydrogene.[29]

Davy made this suggestion quantitative by giving a list of weight relations in which part of the weight of the combustibles was considered as hydrogen in just a sufficient quantity to combine with the known weight of oxygen required for the complete combustion of the material.

> Amongst the acidifiable bodies, sulphur, which is represented by 30, may be supposed to consist of 6 hydrogene, and 24 basis; phosphorus of 4 hydrogene and 16 basis; and charcoal 4 hydrogene and 7.4 basis. It will be unnecessary to supply any more of these estimations, the principles of which are obvious; and in an elementary book it would be improper to dwell upon matters of mere speculation; even these transient views have been developed merely for the sake of pointing out a promising path of enquiry.[30]

This interpretation, though ingenious, proved fruitless and Davy did not pursue the idea further.

In America, Mitchill and Coxe were most enthusiastic in their reception of Davy's work, but there were other men in this country who were more reluctant to accept the phlogistic interpretation of it. One of these was John Maclean, who had earlier taken issue with Priestley. In a letter to Benjamin Silliman dated 10 February 1810, he wrote:

> Have you seen the last number of the *Medical Repository*? If not, be advised that the great Gas Holder Mitchill, in order to excite a blaze, has thrown two bubbles of hydrogene gas at the heads of the antiphlogistians — God help him; like all other tail bearers of Stahl, he thinks he must be right, if his opponents in any one instance have been wrong. I hope that some one will give him, if he is worthy of notice, that flagellation that he merits.[31]

A more outspoken critic of the revived phlogiston concept was Thomas D. Mitchell, M.D. [32]

In 1811, Mitchell wrote a paper called "Remarks on the phlogistic and anti-phlogistic systems of chemistry," [33] in which he first gave a brief history of the two opposing theories, while showing a decided favoritism for the anti-phlogistic side. He recognized the potential significance of Davy's work in supporting the revival of phlogiston, for he stated:

> Notwithstanding the apparent death blow that the Phlogistic system received [at the hands of Lavoisier], there appears some probability of a renewal of the old dispute. The various researches of Davy have given rise to doubts respecting the tenable nature of the whole Antiphlogistic system.[34]

[29] *Loc. cit.*, p. 481.

[30] *Ibid.*, pp. 482–483.

[31] Quoted by William Foster in his edition of Maclean's *Lectures on Combustion*, page 10, from a letter in the Princeton University Library.

[32] Thomas D. Mitchell's opposition to the revival of phlogiston is the more interesting in view of his schooling. He attended the University of Pennsylvania Medical School (graduating in 1812) during the years when John Redman Coxe was professor of chemistry. Any influence that Coxe may have had on Mitchell on this aspect of chemistry must have been negative.

[33] *Memoirs* of the Columbian Chemical Society of Philadelphia, 1813, *1*: 5–14.

[34] *Loc. cit.*, p. 6.

After elaborating on the evidence favoring a revival of phlogiston, he expressed his own views in the following words:

> Much has been said relative to a principle of inflammability in inflammable bodies; but admitting its existence as a possible thing, I see no necessity for it . . . and I believe that all important phenomena can be accounted for independent of such an agent.[35]

In slightly later writings Mitchell was sharply critical of Davy.

> Mr. Davy . . . has appeared the champion of the Phlogistic system and has endeavored to discover hydrogen in almost everything.[36]

> The confusion, in which the Science of Chemistry has been involved, by the researches of Professor Davy, is truly alarming.[37]

Mitchell, in his critical enthusiasm, also questioned Davy's evidence for the elementary nature of oxy-muriatic acid (chlorine). He was hardly fair in calling Davy the "champion of the Phlogistic system," for Davy himself never subscribed to it, but there is no question that Davy did provide a great deal of evidence that such champions needed. And there is some justice in saying that Davy attempted "to discover hydrogen in almost everything," for Davy long held the notion that nitrogen would prove to be a compound of hydrogen and oxygen. Davy had also suggested the possibility that metals were compounds of hydrogen, a view also held by Gay-Lussac and Thénard.

Mitchell's criticism was almost entirely philosophic, and he offered no experimental evidence of his own to support his stand. In reviewing Coxe's "Essay on Combustion," the nature of his criticism is clearly indicated.

> When we speak of the properties of bodies, as taste, smell, etc., we do not mean that any of them possess a positive quality. They are merely sensations or effects resulting from the action of these bodies on our organs of taste, smell, etc. Inflammation, like odors, is the result of relative circumstances and not the product of a single agent.[38]

A little later in the same article Mitchell said:

> With regard to an inflammable principle, I think [it] proper to say that there is as little reason for retaining it as the principle of acidity. Hydrogen as well as oxygen would seem according to the professor, to be a *sine qua non* of combustion, as well as acidity. But pray, why not call the combustible itself a *sine qua non* of combustion? for I believe no one supposes the process can go on without it. Matter, we know has a capacity to be acted upon, but [is] not a principle of action. An alkali has a capacity of being converted into a neutral salt, by union with an acid, but it contains no principle of a neutral salt; and with as much logic may it be said, that a combustible contains no principle of combustion, or inflammability in itself. What is a neutral salt, but the result of the mutual action of an acid and an alkali, and what is combustion, but the effect of the mutual operation of oxygen gas, in some shape or other, and a combustible? [39]

This is a very modern-sounding criticism and a very good one. However,

[35] *Ibid.*, pp. 11–12.

[36] On muriatic and oxy-muriatic acids, combustion, etc., *Memoirs* of the Columbian Chemical Society, 1813, *1*: 103.

[37] Critical remarks on the Bakerian Lecture of 1809, *Medical Repository*, 1812, *15*: 287.

[38] Analysis of Professor Coxe's Essay on Combustion and Acidification, *Memoirs* of the Columbian Chemical Society, 1813, *1*: 179.

[39] *Ibid.*, p. 188.

Mitchell dismissed the actual physical evidence of Davy rather out of hand. Though he was ultimately shown to be correct in doing so, he had no experimental justification for doing so at that time. Davy's evidence could be rationally interpreted as Mitchill and Coxe interpreted it, and indeed Davy himself suggested such a scheme. After all, the argument could only be resolved by experimental evidence and not by logic, no matter how cogent it may seem in retrospect. None the less, the criticism of Thomas D. Mitchell appears refreshingly clear amidst the fog of principles and essences which enveloped the phlogiston controversy generally.

The position of Samuel L. Mitchill in the history of American chemistry has not previously been adequately determined. The customary view has been that he was an early and influential advocate of the antiphlogistic philosophy, while his modifications of the chemical nomenclature after 1797 have been dismissed as terminological detail. His only extensive biographer, Courtney Robert Hall, made the following statement in summary of Mitchill's part in the phlogiston controversy.

> It is somewhat to be regretted that Mitchill did not take a stronger attitude in endorsement of the conclusions of the French chemists. His own beliefs, except in some details of terminology, were Lavoisierian and antiphlogistic.[40]

Mitchill's comments on phlogiston after 1810 were apparently unknown to Hall, for, as presented in this paper, they make it clear that Mitchill's views were definitely not antiphlogistic. That he recognized the role of oxygen in combustion is true, but it is equally true that he insisted upon a principle of inflammability with more enthusiasm than the experimental evidence could justify. Mitchill, by insisting upon a principle of inflammability, required the invention of a large number of even more abstract concepts. Each inflammable, including the metals, was to be considered as composed of phlogiston (hydrogen) plus an as yet undiscovered basis, for which there was no direct evidence of any kind. It is apparent that he was not in full sympathy with the philosophy of the new chemistry of Lavoisier, for he missed the real significance of Lavoisier's work; namely, the establishment of a science of chemistry based on material elements empirically determined.

After the brief flurry of papers around 1810, neither Coxe nor Mitchill published any further comments on phlogiston and its revival. No clear-cut experimental evidence appeared to settle the issue unequivocably, and as late as 1816, John Gorham, Erving Professor of Chemistry at Harvard, could say:

> Hydrogen has been discovered to constitute a part of all bodies, except metals, and should they be proved to contain it, the doctrine of phlogiston must be renewed in a demonstrable and permanent form.[41]

However, confirmation of the presence of hydrogen in metals and other combustibles did not appear, and the attempted revival of phlogiston failed. Without the specific evidence to substantiate it, the phlogiston concept became simply unnecessary and quietly disappeared from the field of chemical interests.

[40] *A Scientist in the Early Republic, Samuel Latham Mitchill, 1764–1831*, New York, 1934, p. 30.

[41] *New England Journal of Medicine and Surgery*, 1817, *6*: 13.

William Charles Wells and the Races of Man

*By Kentwood D. Wells**

IN 1866, while preparing the fourth edition of the *Origin of Species* for publication, Charles Darwin added to his "Historical Sketch" a brief notice of a paper written by William Charles Wells (1757–1817) in 1813. "In this paper," Darwin wrote, "he distinctly recognizes the principle of natural selection, and this is the first recognition which has been indicated." Darwin tempered his praise for Wells with the remark that "he applies it [natural selection] only to the races of man, and to certain characters alone."[1]

Despite Darwin's statement of the importance of Wells' ideas, relatively little attention has been given to Wells' place in the history of science. Most authors who have discussed him at all have treated him as a forerunner of Darwin who stumbled upon a great truth but failed to fully recognize the significance of his discovery.[2] He is generally considered to have been a man whose ideas came before their time and consequently fell on deaf ears. Great emphasis has been placed on the similarity of his ideas to those of Charles Darwin. Yet few authors have examined the relationship between Wells' ideas and those of his contemporaries. This paper is an attempt to re-evaluate Wells' views on natural selection and race formation from this point of view.

William Wells was born in Charleston, South Carolina, on May 24, 1757. In 1770 he travelled to Edinburgh to study at the university there, but he returned to Charleston the next year to begin an apprenticeship with Dr. Alexander Garden. Wells' family was sympathetic to the British government during the American Revolution, and he moved to Edinburgh in 1775 to begin medical studies. He later settled in London and became a student at St. Bartholomew's Hospital. After receiving an M.D. degree in 1780, Wells returned to the United States for four years, but in 1785 he settled permanently in London.[3]

Received Sept. 1971: revised/accepted Apr. 1972.

* Section of Ecology and Systematics, Cornell University, Ithaca, New York 14850. I would like to express my appreciation to Dr. William B. Provine, Department of History, Cornell University, for reading the manuscript and making numerous helpful suggestions. Dr. K. A. R. Kennedy, of the Department of Anthropology, also read parts of the manuscript. (The author informs us that he is, regretfully, unrelated to his subject. ED.)

[1] Charles Darwin, *The Origin of Species: A Variorum Text*, ed. Morse Peckham (Philadelphia: University of Pennsylvania Press, 1959), p. 61.

[2] Richard Harrison Shryock, "The Strange Case of Wells's Theory of Natural Selection (1813): Some Comments on the Dissemination of Scientific Ideas," in *Studies and Essays in the History of Science and Learning Offered in Homage to George Sarton*, ed. M. F. Ashley Montagu (New York: Henry Schuman, 1944), p. 205.

[3] Norman Moore, "William Charles Wells," *Dictionary of National Biography*, Vol. XX, pp. 1146–1147.

The scant biographical information available on Wells fails to reveal anything particularly notable in his life.[4] He published a number of minor papers in the *Philosophical Transactions of the Royal Society* which have long since been forgotten. In 1792 he published his *Essay on Single Vision With Two Eyes,* which was widely read and admired at the time. In 1814 he published his *Essay on Dew,* for which he received the coveted Rumford Medal of the Royal Society.[5] Yet it is a single paper, a mere twelve pages long, for which Wells is remembered today.

Wells' paper was first read to a meeting of the Royal Society in 1813. It dealt with observations he had made on a piebald woman named Hannah West. He used these observations as a base for speculations on the origin of human racial differences. The paper was subsequently published as an appendix to a new edition of his *Two Essays: One on Dew and the Other on Single Vision With Two Eyes* (1818), with the improbable title, "An Account of a Female of the White Race of Mankind, Part of Whose Skin Resembles that of a Negro; with Some Observations on the Causes of the Differences in Colour and Form Between the White and Negro Races of Men."[6]

Wells noted that different races of men were susceptible to different diseases. Thus Europeans in tropical Africa often died from local diseases, while the native population was less affected. The same was true of Africans transferred to northern latitudes. Wells therefore concluded that colonies of Europeans in Africa would tend to die out or become weak and feeble. He believed that the differences in color among races, while not the cause of their varying susceptibility to diseases, was "a sign of some difference in them."[7] He then went on to show how this adaptation to the diseases and climate of a certain region could account for the formation of human races. Noting that "among men, as well as among other animals, varieties of a greater or less magnitude are constantly occurring," Wells stated that man was able to select these variations to improve his domesticated animals. He continued,

> But what is here done by art, seems to be done, with equal efficacy, though more slowly, by nature, in the formation of varieties of man, fitted for the country which they inhabit. Of the accidental varieties of man, which would occur among the first few and scattered inhabitants of the middle regions of Africa, some one would be better than the others to bear the diseases of the country. This race would consequently multiply, while the others would decrease, not only from their inability to sustain the attacks of disease, but from their incapacity of contending with their more vigorous neighbors. The colour of this vigorous race I take for granted, from what has been already said, would be dark. But the same disposition to form varieties still existing, a darker and darker race would in the course of time occur, and as the darkest would be best fitted for the climate, this would at length become the most prevalent, if not the only race, in the particular country in which it had originated.[8]

Wells believed that the influence of this natural selection on the color of the human race "would necessarily operate chiefly during its infancy," when men were "a few

[4] See F. L. Pleadwell, "That Remarkable Philosopher and Physician, Wells of Charleston," *Annals of Medical History,* 1934, *4*:128–149, for some anecdotes of Wells' life.

[5] Moore, "William Wells," p. 1147.

[6] William Charles Wells, *Two Essays, One on Dew and the Other on Single Vision With Two Eyes* (Edinburgh: Archibald and Constable, 1818). A reprint of Wells' paper is available in H. Lewis McKinney, *Lamarck to Darwin: Contributions to Evolutionary Biology 1809–1859* (Lawrence, Kansas: Coronado Press, 1971), pp. 21–28.

[7] Wells, *Two Essays,* p. 434.

[8] *Ibid.,* pp. 434–436.

wandering savages." Once men had adopted a "more refined mode of life," they would tend to retain certain customs and practices that would preserve the original race.[9]

Wells' paper attracted very little attention when it first appeared, although I have located two references to it which are worth noting. Alexander Tilloch, who was at that time the editor of *The Philosophical Magazine*, made a point of attending meetings of various scientific societies and reporting on the papers which were presented. In his report of the meeting of the Royal Society on April 1, 1813, Tilloch noted that "Dr. C. Wells communicated an account of Harriet Trest, a woman who has her left shoulder, arm, and hand as black as the blackest African, while all the rest of her skin is very white." He went on to state Wells' conclusion that "blackness of skin is no proof of differences of species and that the sun does not blacken but rather whitens the skin."[10]

The theoretical portion of Wells' paper was presented on April 8, and Tilloch reported it as follows: "The doctor indulged a variety of speculations; supposed with Volney, that the Egyptians were negroes; conceived that the want of civilization contributes to make people black; and referred to various South-sea islanders and others, to sanction this singular fancy." Tilloch added, with obvious annoyance, that "the length of these conjectures prevented the reading of a valuable paper by Professor Berzelius and Dr. Marcet."[11]

Wells' paper received a more sympathetic and detailed treatment in Thomas Thomson's *Annals of Philosophy*. While Tilloch had managed to miss the whole natural selection argument, Thomson gave the essence of Wells' theory in his summary:

> On Thursday the 8th of March [April] Dr. Wells's paper was concluded. He gave his opinion about what occasioned the difference between negroes and whites. It is well known that whites are not so well able to bear a warm climate as negroes; and that they are liable to many diseases in such a situation, from which negroes are free. On the other hand, whites are much better fitted to bear a cold climate than negroes. Suppose a colony of whites transported to the torrid zone, and obliged to subsist by their labour, it is obvious that a great proportion of them would speedily be destroyed by the climate and the colony, in no long period of time, annihilated. The same thing would happen to a colony of negroes transported to a cold climate. Dr. Wells conceives that the black colour of negroes is not the cause of their being better able to bear a warm climate, but merely the sign of some difference in constitution which makes them able to bear such a climate. Suppose a colony of white men carried to the torrid zone; some would be better able to resist the climate than others. Such families would thrive, while the others decayed. These families would exhibit the sign of such a constitution; that is, they would be dark: and as the darker they were, the better they would be able to resist the climate; it is obvious, that the darker varieties would be the most thriving, and that the colony, on that account, would become gradually darker and darker coloured till they degenerated into negroes. The contrary would happen to negroes transported to cold climates.[12]

Here, in a journal published in 1813, is a theory of race formation by natural selection. Yet I have found no firm evidence that this account attracted any attention, just as the paper attracted no attention when published in Wells' widely read *Two Essays* in 1818. In fact, until Darwin referred to the paper in his "Historical Sketch" in 1866, it remained lost in obscurity.

[9] *Ibid.*, p. 436.

[10] [Alexander Tilloch], "Proceedings of Learned Societies," *The Philosophical Magazine*, 1813, *41*:302.

[11] *Ibid.*, pp. 302–303.

[12] [Thomas Thomson], "Proceedings of Philosophical Societies," *Annals of Philosophy*, May 1813, *1*:383.

There has been some discussion of the reasons for Wells' failure to attract any attention to his theory. There has also been considerable controversy over the question of how completely Wells anticipated the Darwin-Wallace view of natural selection. As we shall see, these two problems are closely related. Shryock has argued that Wells "put together two hitherto isolated ideas which had long been familiar to biologists—'natural selection' and the 'origin of species'—in an essentially new combination."[13] Although Wells emphasized the application of his theory to human races, Shryock believed that "Wells plainly envisaged the theory as applicable to animal forms."[14] Similarly Conway Zirkle, in his study of the history of natural selection, states that Wells "obviously understood its wider application."[15] Loren Eiseley has been more cautious in his interpretation. While acknowledging that Wells mentioned animals briefly and clearly anticipated natural selection, Eiseley makes the important observation that there is no evidence that Wells grasped the concept of unlimited organic change in time.[16] Thus it is not at all clear that Wells attributed the origin of new species to natural selection, as Shryock maintained.

In my own view, I believe that Wells not only failed to grasp the idea of species changes through time, but he probably did not intend to apply the process of natural selection to wild animals. He merely stated that "among men, as well as among other animals, varieties of a greater or less magnitude are constantly occurring."[17] He further stated that man could select these varieties in domesticated animals. The only mention of selection "by nature" referred to the formation of human races or varieties.

Thus it is inappropriate to group Wells with men such as Erasmus Darwin, Lamarck, and Robert Chambers, who were concerned with the evolution of species through time. Instead, he belongs with the group of physician-anthropologists whose works were appearing in the first two decades of the nineteenth century. During this period such British physicians as James Cowles Prichard (1786–1848) and William Lawrence (1783–1867) were engaged in speculation on the origin of human races. In the United States men like Samuel Stanhope Smith and a host of later writers also produced treatises explaining the origin of racial differences.[18]

The early years of the nineteenth century marked the emergence of anthropology as a science in its own right. Earlier discussions of human races had been largely speculative, with very little support from actual observations. The great eighteenth-century naturalists, Linnaeus and Buffon, had attempted to classify races according to physical characteristics. Anatomists like Petrus Camper (1722–1789) and John Hunter (1728–1793) had assembled and studied collections of human skulls and quantified physical differences between races. However, it was only in the work of the great German

[13] Shryock, "Strange Case," p. 203.

[14] *Ibid.*

[15] Conway Zirkle, "Natural Selection Before the 'Origin of Species,'" *Proceedings of the American Philosophical Society*, 1941, *84*:106.

[16] Loren Eiseley, *Darwin's Century* (New York: Doubleday, 1959), p. 122. John C. Greene, *The Death of Adam* (Ames, Iowa: Iowa State University Press, 1959), pp. 244–247, also discusses Wells' theory, but he does not speculate on whether Wells understood the wider applications.

[17] Wells, *Two Essays*, p. 434.

[18] On Prichard and Lawrence, see Kentwood D. Wells, "Sir William Lawrence (1783–1867): A Study of Pre-Darwinian Ideas on Heredity and Variation," *Journal of the History of Biology*, 1971, *4*:319–361; on Smith, see Samuel Stanhope Smith, *An Essay on the Causes of the Variety of Complexion and Figure in the Human Species*, ed. Winthrop D. Jordan (Cambridge, Mass.: Harvard University Press, 1965; reprint of 2nd ed., New Brunswick, N.J., 1810).

anatomist Johann Friedrich Blumenbach (1752–1840) that a cohesive science of anthropology began to appear.[19]

In England in 1813 the great debate over the unity versus plurality of the human species, which dominated anthropology at mid-century, had scarcely begun. A few writers, notably Lord Kames (1696–1782) and Charles White (1728–1813), had championed the polygenist doctrine which held that each human race constituted a separate species of man.[20] For the most part, however, the monogenists had the field to themselves. This was due mainly to the influence of Blumenbach, who was the intellectual godfather of the two most important anthropologist-ethnologists of the day, James Cowles Prichard and Sir William Lawrence. These men, following Blumenbach's lead, stated as their basic premise the fact that all human races constituted a single species. Yet Blumenbach, Prichard, and Lawrence could scarcely deny that there were enormous physical differences among the races of man, for these differences were readily observable in nature. It is not surprising, therefore, that much of their attention became focused on the problem of explaining the origin of human racial variations.

In the late eighteenth century Blumenbach made a very significant contribution to the study of variation when he suggested that variations in the human species might be explained by examining analogous variations in domesticated animals. His aim was to show that the differences between human races were no greater than the differences between varieties of a single species of domesticated animal. To prove his point Blumenbach cited numerous examples from both the human and animal worlds, thereby assembling a valuable compendium of information which would serve as a base for future workers.[21]

Prichard's *Researches Into the Physical History of Man* first appeared in 1813, although a shorter version had been published in Latin as a dissertation in 1808.[22] In 1819 Lawrence's *Lectures on Physiology, Zoology, and the Natural History of Man* appeared in print.[23] Prichard's and Lawrence's ideas were very similar, and Lawrence's work owed much to the 1813 edition of Prichard's book.[24] Both men agreed with Blumenbach that variations within the human species were no greater than those within domesticated animals, and they cited many of his examples. However, when they came to consider the causes of these variations, Prichard and Lawrence diverged sharply from their German colleague.

Blumenbach was very traditional in his explanation of the causes of variations.

[19] This early period of anthropology has not been well studied, but a few good sources are available. For an excellent discussion of the prescientific period, see Margaret T. Hodgen, *Early Anthropology in the Sixteenth and Seventeenth Centuries* (Philadelphia: University of Pennsylvania Press, 1964); see also D. J. Cunningham, "Anthropology in the Eighteenth Century," *Journal of the Royal Anthropological Society*, 1908, *38*: 10–35; William Stanton, *The Leopard's Spots: Scientific Attitudes Toward Race in America 1815–59* (Chicago: University of Chicago Press, 1960); on Blumenbach, see Thomas Bendyshe, ed., *The Anthropological Treatises of Johann Friedrich Blumenbach* (London: The Anthropological Society, 1865).

[20] Stanton, *Leopard's Spots*, pp. 15–18.

[21] Bendyshe, *Anthropological Treatises of Blumenbach, passim.*

[22] James Cowles Prichard, *Disputatio inauguralis de generis humani varietate* (Edinburgh: Abernethy and Walker, 1808); *Researches Into the Physical History of Man* (London: J. and A. Arch, 1813).

[23] I have used the following edition: William Lawrence, *Lectures on Physiology, Zoology, and the Natural History of Man* (Salem, Mass.: Foote and Brown, 1828).

[24] I have discussed the relationship between Lawrence and Prichard more fully in "Sir William Lawrence," pp. 351–361.

Environmental factors such as climate, food, and way of life were assumed to be capable of producing inheritable changes in the physical constitution of men and animals. Even artificial constriction of the head and other such man-made changes were thought to produce inheritable effects. In Blumenbach's theory these external factors acted gradually on a whole race to produce modifications in skin color or hair texture. Thus, a light-colored race would become gradually darkened after residing in a hot tropical environment.[25] This was simply a reiteration of the time-honored theory of the inheritance of acquired characteristics, which, as Zirkle has shown, had been generally accepted for centuries.[26]

Prichard and Lawrence, on the other hand, emphatically rejected this ancient theory. The reasons for this are unclear, but they were unequivocal in stating their views.[27] "There is no foundation," Prichard wrote, "for the common opinion which supposes the black races of men to have acquired their colour by exposure to the heat of a tropical climate during many ages. . . . The offspring . . . inherit only their connate peculiarities, and not any of the acquired qualities."[28] Similarly, Lawrence declared, "In all the changes which are produced in the bodies of animals by the action of external causes, the effect terminates in the individual; the offspring is not in the slightest degree modified by them."[29]

Prichard and Lawrence suggested that the differences among human races originated from hereditary variations which appeared spontaneously in certain individuals and which were perpetuated through breeding. Thus, while Blumenbach had emphasized environmental factors which affected whole races, Prichard and Lawrence shifted the emphasis to individual variations. As Prichard expressed it,

> The children of the same parents, though often bearing a general resemblance yet exhibit always some difference and frequently a considerable diversity in these respects. . . . By observing that such a tendency to deviation exists even among the individuals of the same family, and that whatever examples of variations may arise, have a general disposition to become hereditary, we appear to make some progress towards an explanation of the diversities of figure, which characterize different races of the human kind.[30]

In his thinking on the origin of variations Wells was in essential agreement with Prichard and Lawrence, although it seems unlikely that any direct influence was involved.[31] Wells stated that black skin color was not a characteristic of a distinct species of man. Furthermore, he rejected environmentalist explanations of the origin of variations and denied that the sun could produce a blackening of the skin.[32] Like Prichard and Lawrence, Wells utilized the analogy between man and domesticated

[25] Bendyshe, *Anthropological Treatises of Blumenbach*, pp. 129, 196–200.

[26] Conway Zirkle, "The Early History of the Idea of Acquired Characteristics and of Pangenesis," *Transactions of the American Philosophical Society*, 1946, N.S. *35*:91–151.

[27] Lawrence accepted the inheritance of acquired characteristics in writings which appeared prior to his 1819 lectures. He seems to have changed his views as a result of the influence of Prichard. See K. Wells, "Sir William Lawrence," pp. 351–361.

[28] Prichard, *Researches* (1813), p. 230.

[29] Lawrence, *Lectures*, p. 436.

[30] Prichard, *Researches* (1813), p. 34.

[31] Wells could not have read Lawrence's 1819 *Lectures* before delivering his own paper in 1813. Wells' paper was given in April 1813, while the preface to Prichard's *Researches* is dated Nov. 1813. Wells could have read Prichard's 1808 dissertation (see n. 22 above), but Wells makes no references to any of his sources. There seems not to have been any influence of Wells on Lawrence, although Prichard may later have read his paper (see n. 37 below).

[32] Wells, *Two Essays*, p. 431.

animals, noting that spontaneous variations "of a greater or less magnitude are constantly occurring" among them.[33]

When he came to a consideration of the mechanism by which such spontaneous variations were preserved in a race, Wells diverged from Prichard and Lawrence. These men noted that man was capable of preserving spontaneous variations in domesticated animals through selection, and they suggested that such principles could be applied to humans as well. They also noted that different ideas of beauty among races might act as a sort of sexual selection and exert an influence on the physical appearance of races. In addition, they were aware of the fact that individuals with similar characteristics might become isolated and form a new race through interbreeding.[34]

Wells also referred to man's ability to perpetuate variations in domesticated animals through selective breeding. However, previous quotations from Wells' work have shown that he clearly understood that a natural process of selection could occur in the formation of human races. In formulating this theory Wells demonstrated a grasp of the concept of adaptation which was notably lacking in the works of Prichard and Lawrence. The survival of a race depended on its ability to adapt to the diseases of the region which it inhabited. Wells was thus able to present a plausible explanation for the distribution of different racial characters as they exist in the modern world. Prichard and Lawrence, while suggesting ways in which variations could be preserved in humans, admitted that they could not satisfactorily explain why particular racial characters had been perpetuated. In this respect Wells' ideas were more "advanced" than those of his contemporaries.

While many of the ideas on variation expressed by these men, particularly Wells, seem remarkably modern to us today, we must remember that all of these anthropologists were concerned with explaining variations within a single species, *Homo sapiens*, and not with the evolution of new species.[35] Current speculations on evolution, as expressed in the writings of Lamarck, Erasmus Darwin, and others, were never mentioned by Prichard, Lawrence, and Wells. It is therefore not surprising that someone like Wells failed to formulate a full-fledged anticipation of Darwinism, for he was functioning in an intellectual world completely different from that of the evolutionists.

It is apparent that the question of why Wells' theory failed to attract much attention is essentially an irrelevant one. This question assumes that his theory would have been recognized by his contemporaries as something bold and innovative. Yet, as I have attempted to demonstrate, Wells' ideas were basically in the mainstream of current anthropological thinking on the nature of variation. True, he had formulated a theory of natural selection which was in advance of the ideas of his contemporaries, but even in this he was not unique, since he did not apply it to the origin of new species. Various concepts of natural selection had been advanced before.[36] Even Prichard, in the

[33] *Ibid.*, p. 434.

[34] Prichard, *Researches* (1813), pp. 38–44; Lawrence, *Lectures*, pp. 387–395.

[35] Various authors have cited Prichard and Lawrence, in addition to Wells, as forerunners of Darwin. See esp. C. D. Darlington, *Darwin's Place in History* (New York: Macmillan, 1955), pp. 16–24; Philip G. Fothergill, *Historical Aspects of Organic Evolution* (London: Hollis and Carter, 1952), pp. 83–86; E. B. Poulton, "A Remarkable Anticipation of Modern Views on Evolution," in *Essays on Evolution* (Oxford: Clarendon Press, 1908), pp. 173–192.

[36] Zirkle, "Natural Selection Before the 'Origin,'" pp. 71–123.

second edition of his great anthropological treatise, expressed ideas remarkably similar to those of Wells.[37] Yet Prichard continued to be concerned only with variations within the human species.

Much of the inclination of historians to read anticipations of Darwinian evolution into the thoughts of earlier writers stems from a failure to consider carefully the ways in which such scientific theories come into being. In addition, there is the constant problem of hindsight colored by an intellectual world in which evolution is taken for granted. While the reality of the *process* of evolution by natural selection has been accepted by modern science, we must bear in mind that the *concept* of natural selection did not exist independently as a sort of Platonic ideal, waiting to be discovered, recognized, or stumbled upon by a Wells or a Darwin or a Wallace. Both Darwin and Wallace had fully accepted the reality of evolution when they formulated the concept of natural selection to provide a mechanism for the origin of new species. Wells was not even thinking in terms of evolution when he hit upon natural selection as a mechanism for preserving human racial variations. To expect him to have anticipated Darwinism, by accident or design, would therefore be unreasonable. It is not that he failed to find the correct answers; he simply had not asked the same questions.

The precise way in which Darwin became aware of Wells' paper has never been fully explained. The story is worth telling, for it provides an interesting sidelight on this episode in the history of ideas. In the fourth edition of the *Origin* Darwin stated, "I am indebted to the Rev. Mr. Brace, of the United States, for having called my attention to the above passage in Dr. Wells' work."[38] In the fifth edition, published in 1869, Darwin modified his acknowledgement to read: "I am indebted to Mr. Rowley, of the United States, for having called my attention, through Mr. Brace, to the above passage in Dr. Wells' work."[39]

Shryock has suggested that "Mr. Brace" may have been Charles Loring Brace, the American clergyman. He noted that Mr. Rowley's identity is unknown, but that "if something were known of the latter's familiarity with Wells's theory, it might throw light on its possible—even if obscure—influence on the biological thought of the mid-nineteenth century."[40]

Darwin apparently first learned of Wells' paper sometime in 1865, when he mentioned in a letter to Joseph Dalton Hooker that "a Yankee" had referred him to it.[41] This "Yankee" was presumably either Brace or Rowley. The "Rev. Mr. Brace" Darwin mentioned in his "Historical Sketch" was undoubtedly Charles Loring Brace,

[37] Prichard, in the 2nd ed. of his *Researches into the Physical History of Mankind*, 2 vols. (London: J. and A. Arch, 1826), recognized a sort of natural selection very similar to the theory advanced by Wells. He postulated that races not adapted to the diseases of a particular region would die out. Prichard may have known of Wells' paper, but I have found no definite evidence to prove that he read it. However, he did read another article in the same volume of the *Annals of Philosophy* which contained the summary of Wells' theory, and he cited it in his book. See Prichard, *Researches* (1826), Vol. II, pp. 550, 573–574, 581.

[38] Darwin, *Origin* (Variorum Text), p. 62.

[39] *Ibid.*

[40] Shryock, "Strange Case," p. 199.

[41] Francis Darwin, ed., *The Life and Letters of Charles Darwin* (London: John Murray, 1887), Vol. III, p. 41. Both Shryock ("Strange Case," p. 199) and Eiseley (*Darwin's Century*, p. 125) incorrectly assume that Darwin learned of Wells' paper in 1860, when the "Historical Sketch" was first written. Morse Peckham's variorum edition of the *Origin* has shown that changes were made in the "Historical Sketch" as well as in the text and that the reference to Wells did not appear until 1866.

as Shryock has suggested, but who was "Mr. Rowley" and how did he come to know of Wells' work?

The most likely answer comes from an article which appeared in the *American Journal of Science and Arts* for September 1868. Written by one Samuel Rowley, it was entitled "A New Theory of Vision."[42] In it are several references to Wells' *Essay Upon Single Vision with Two Eyes*, in the London 1818 edition, the one which contained the appendix on race formation. Unfortunately nothing more is known about Rowley, and I have found nothing else written by him. It seems likely, however, that this is Darwin's "Mr. Rowley, of the United States." Apparently Rowley happened upon the natural selection paper while reading the essay on vision, in preparation for his article, and recognized that it anticipated Darwin's views.

However, Darwin mentioned that he learned of the paper from Mr. Rowley "through Mr. Brace." Did Rowley first send word of the Wells paper to Charles Loring Brace, a clergyman, and if so, why? It seems that Brace was something of an amateur anthropologist, and in 1863 he published a book entitled *The Races of the Old World: A Manual of Ethnology*.[43] What is more, he made a point of praising Darwinian natural selection, and he applied natural selection to the formation of human races.

Brace theorized that in "some very remote age of the past" a tribe might have moved from Asia or elsewhere to eastern Africa. The climate and diseases would be different from those in the tribe's native region. In some offspring slight varieties might appear which fitted "the possessors to resist the destructive influences of the new climate," and these would be more likely to survive. Therefore Brace saw "no difficulty on this supposition—on the Darwinian theory of an imperceptible accumulation of profitable change through long periods of time . . . of accounting for the origin of the negro from the white man, or from the brown, or from some other race."[44]

We can see that Brace was quite clear in his application of natural selection to human races. It therefore seems possible that Rowley called Brace's attention to the Wells paper for his *own* information rather than for Darwin's. Yet one question remains: How did Rowley happen to know Brace? It turns out that they were neighbors! The preface to Brace's book was written at his home in "Hastings-on-the-Hudson" in April 1863.[45] Fortunately for later historians, Rowley included his own address with his article: "Hastings-upon-Hudson, N.Y."[46]

The story of Wells' theory does not end with Darwin's brief mention of it in his "Historical Sketch." Darwin himself had independently developed a mechanism similar to Wells' whereby "negroes and other dark races might have acquired their dark traits by the darker individuals escaping from the deadly influence of the miasma of their native countries during a long series of generations."[47] As early as 1862 he even hit upon the rather novel (although for Darwin, typical) idea of sending questionnaires to Army doctors in all parts of the British Empire inquiring about the different susceptibilities of the British and native soldiers to diseases.[48] Darwin told Wallace of

[42] Samuel Rowley, "A New Theory of Vision," *American Journal of Science and Arts*, 1868, 2nd ser., *46*:153–167.

[43] Charles Loring Brace, *The Races of the Old World: A Manual of Ethnology* (New York: Scribner's, 1863).

[44] *Ibid.*, pp. 496–499.

[45] *Ibid.*, p. iv.

[46] Rowley, "New Theory of Vision," p. 167.

[47] Charles Darwin, *The Descent of Man* (2nd ed.; New York: Hurst and Co., 1874), p. 209.

[48] *Ibid.*, p. 210n.

his theory and questionnaire in a letter written in May 1864.[49] In his reply Wallace stated that while that was all very interesting, he did not expect the results of the inquiry "to be favourable to your view."[50] This is not surprising, since in his 1864 paper on human races and natural selection, which he and Darwin were discussing in these letters, Wallace declared that man's physical constitution was no longer influenced by natural selection.[51] As it happened, Darwin got no replies from his inquiry and later concluded that there was no evidence for his theory.[52]

When Wallace read about Wells in Darwin's "Historical Sketch" he wrote to Darwin, "How curious it is that Dr. Wells should so clearly have seen the principle of Natural Selection fifty years ago, and that it should have struck no one that it was a great principle of universal application in Nature!"[53] A few years later Wallace brought out his *Contributions to the Theory of Natural Selection* (1870), and he mentioned Wells in the preface. The 1864 paper was republished under the title "The Development of the Human Races Under the Law of Natural Selection." Curiously enough, Wallace added one paragraph to the original paper:

> There is one point, however, in which nature will still act upon him as it does on animals, and to some extent, modify his external characters. Mr. Darwin has shown that the colour of the skin is correlated with constitutional peculiarities both in vegetables and animals, so that liability to certain diseases or freedom from them is often accompanied by marked external characters. Now there is every reason to believe that this has acted, and, to some extent, may still continue to act, on man. In localities where certain diseases are prevalent, those individuals of savage races which were subject to them would rapidly die off; while those who were constitutionally free from the disease would survive, and form the progenitors of a new race. These favoured individuals would probably be distinguished by peculiarities of *colour*, with which again peculiarities in the texture or the abundance of *hair* seem to be correlated, and thus may have been brought about those racial differences of *colour*, which seem to have no relation to mere temperature or other obvious peculiarities of climate.[54]

This paragraph was a direct contradiction of the main point of Wallace's paper as originally written. Although the evidence is circumstantial, it seems very likely that Wallace was prompted to add this paragraph after reading Darwin's account of Wells' ideas. Apparently he found Wells' arguments of 1813 so convincing that he decided to make an exception to his belief that natural selection could no longer act on man's body.[55]

[49] James Marchant, ed., *Alfred Russel Wallace: Letters and Reminiscences* (New York: Harpers, 1916), p. 128.

[50] *Ibid.*, p. 129.

[51] Alfred Russel Wallace, "The Origin of Human Races and the Antiquity of Man Deduced From the Theory of Natural Selection," *Journal of the Anthropological Society of London*, 1864, *2*: clxiii–clxv; clxviii–clxix.

[52] Darwin, *Descent*, p. 211.

[53] Marchant, *Wallace*, p. 145.

[54] Alfred Russel Wallace, "The Development of the Human Races Under the Law of Natural Selection," in *Contributions to the Theory of Natural Selection* (London: Macmillan, 1870), p. 316.

[55] Wallace had been interested in the origin of human races since at least 1845, when he read the works of Prichard and Lawrence and was impressed by their ideas on variation. For additional information on Wallace's anthropological interests, see H. Lewis McKinney, "Alfred Russel Wallace and the Discovery of Natural Selection," *Journal of the History of Medicine and Allied Sciences*, 1961, *21*: 333–357; H. Lewis McKinney, *Wallace and Natural Selection* (New Haven, Conn.: Yale University Press, 1972); K. Wells, "Sir William Lawrence," pp. 339–344.

With Wallace's statement we have come full circle to return to our starting point. Nearly sixty years earlier William Wells had formulated a theory of natural selection without giving any consideration to the evolutionary implications of his ideas. Wallace, for whom evolution and natural selection had long since become established facts, thought he detected the ghost of a fellow evolutionist emerging from the shadows of the past. He sought to revive these long-lost ideas by applying them to one aspect of his own evolutionary philosophy. Ironically, in doing so, Wallace was preparing a solution to the only problem Wells had considered—the origin of human racial variations.

The Quaker Background and Science in Colonial Philadelphia

By Brooke Hindle *

THE extensive thought that has been given to the role played by Quakers in the advance of science has generally been limited to description and to speculation upon the causes of Quaker interest in science. What sort of man was impelled — under the Quaker influence — to encourage scientific activity and sometimes to make contributions of his own to the advance of science has not been made clear. Indeed, the question has hardly been asked.

That the Quaker influence was beneficial to the development of science seems well attested by the numbers of eighteenth-century English and American patrons and scientists who were members of the Society of Friends. Why that was the case has sometimes been explained by purely external factors — by the circumstances which closed all professional careers except medicine to English Quakers. In their medical training, many Quakers gained the best scientific education and were led to varied scientific interests.[1] Yet, however valid this explanation may be for eighteenth-century England, it is not very helpful in the case of Pennsylvania, where Friends were limited only by the tenets of their own faith.

A more fruitful approach is to explain the Quaker interest in science by an extension of the asserted correlation of "the Protestant Ethic" with the encouragement of science.[2] Frederick B. Tolles has demonstrated that Quaker interest in science did follow in a most natural way from the internal structure of Quaker thought. He has pointed out that the Friends shared many of the values of Puritanism. In particular, they favored an empirical, rational approach to knowledge which was in the closest harmony with the pursuit of science.[3] Living in and of the world, unlike some other radical sects, the Quakers found a positive reason for encouraging science in the prevailing expectation that it would ultimately improve the physical condition of man's life.

A fundamental characteristic of the Quaker outlook was concern for things rather than words — a concern that was not merely permissive but positive. "Languages are not to be despised or neglected," William Penn declared, "But things are still to be preferred." Again and again, Penn urged reliance upon "right reason," which to him meant not scholastic "Words and Rules" but

* New York University.

[1] See for example David J. Davis, "The Quakers and Medicine," *Bulletin of the Society of Medical History of Chicago*, 1928, *4*: 80; Arthur Raistrick, *Quakers in Science and Industry* (London, 1950), 221–315.

[2] The phrase is Max Weber's, but its applicability to English Puritanism has been particularly examined: Robert K. Merton, "Science, Technology and Society in Seventeenth Century England," *Osiris*, 1938, *4*:360–632; Dorothy Stimson, "Puritanism and the New Philosophy in the Seventeenth Century," *Bulletin of the Institute of the History of Medicine*, 1935, *3*:321–34; George Rosen, "Left-Wing Puritanism and Science," *ibid.*, 1944, *15*:375–80.

[3] Frederick B. Tolles, *Meeting House and Counting House* (Chapel Hill, 1948), 205–13.

thought grounded in "*Mechanical* and *Physical* or natural *Knowledge*." In his *Works*, his followers read, "The World is certainly a great and stately Volume of natural Things; and may be not improperly stiled the *Hieroglyphicks* of a better: but alas, how very few leaves of it do we seriously turn over! . . . It were happy if we studied *Nature* more in natural Things; and acted *according* to Nature; whose rules are *few*, *plain* and *most reasonable*." [4]

Certainly, the Quaker pattern of thought contained elements that favored the study of science, but the Quaker influence was felt in very different measure by different men. In eighteenth-century Philadelphia, every imaginable degree of Quaker influence was represented. John Woolman brought often to the city the purest sort of almost other-worldly concern with the religious teachings of the Society of Friends. Anthony Benezet devoted his life to humanitarian efforts to improve the life of the unfortunate. Israel Pemberton and Henry Drinker spent their lives extending great fortunes, and yet they remained pillars of the Society. There were still other men in whom influences external to the Society of Friends came to bear increasing sway. Some of this group remained within the Society even though they held more worldly views than their fellow Friends. Some joined other religious bodies. Some had to be disowned by the Society of Friends when their activities or beliefs deviated too widely from the Quaker pattern. With all of these, the Quaker influence remained strong — very strong. It was strong too, among numbers of men who had never been members of the Society of Friends but whose relatives and closest friends were members of the Society and who placed high value upon the Quaker way.

It was precisely these peripheral Quakers, former Quakers, and near-Quakers, who made the most enthusiastic efforts to advance science in the city of Philadelphia. That fact emerges from an examination of the circumstances surrounding the founding of America's first learned society, the American Philosophical Society, held in Philadelphia, for Promoting Useful Knowledge. This body was established in January 1769 by the union of two previously existing societies, each of which had sought to stimulate the advance of science.

One root of the 1769 society went back to 1750 with the founding of an obscure club of young men who met for the purpose of self-improvement through conversation and study. Meeting for years in secret, this "young Junto," as it was called, was revived in 1766 and gradually transformed into a much more important sort of organization. The revival, following a three-year interruption of all meetings, was carried through by three men who were more strongly under the Quaker influence than might at first appear. They were: Charles Thomson, a Presbyterian importer; Edmund Physick, an Anglican civil servant; and Isaac Paschall, a Quaker ironmonger.[5]

The driving force was supplied by Charles Thomson, who, although raised

[4] William Penn, "Some Fruits of Solitude in Reflexions and Maxims," *Collection of the Works of William Penn* (London, 1726), I, 820; William Penn, "The Sandy Foundation Shaken," *Select Works* (London, 1771), 9, 17, 20; some of these passages are also quoted in Tolles, *Meeting House*, 208.

[5] Junto Minutes [1758–1762], American Philosophical Society; American Society Minutes, 25 April 1766, p. 2, American Philosophical Society; Brooke Hindle, The Rise of the American Philosophical Society, 1766 to 1787 (Doctoral Dissertation, University of Pennsylvania, 1949), 33–42, 51–53.

as a Presbyterian, had a peculiar affinity for Quaker friends and a marked respect for the Quaker faith. Before turning his hand to business, he had taught in the Friends' school and when married for a second time on the eve of the Revolution, he chose a Quaker wife. Ultimately, he left the Presbyterian church to take up a position, "attached to no system nor peculiar tenets of any sect or party." [6] He spent many years at the end of his life translating the Bible. Thomson had so many relationships with the Quakers and so many of their characteristics that he has even been mistaken for one.

Edmund Physick was a Proprietary officer who belonged to the political faction which bitterly opposed the aims and aspirations of the Quaker party. Yet even he demonstrated an unusual closeness to the Quaker group in his personal associations and in his decision to send his son to the Friends' school.[7]

Isaac Paschall was a Quaker, as were the remaining six members who constituted the "young Junto" in 1766, but they were Quakers of a very special stamp. Before the Revolution was over, four of these seven men had either been disowned or had voluntarily left the Society of Friends. Three of them took an active part in the conflict and became leaders in the Free Quaker movement. None of the seven was ready to follow the Quaker leaders into the unhappy wartime exile at Winchester, Virginia. These were worldly Quakers, but even when they came to the point of breaking with the Society, they retained the bulk of their Quaker heritage.[8]

As the club grew, its aspirations expanded along the lines of the Royal Society of London and the Society of Arts in London but with a liberal infusion of American patriotism. The name it decided to adopt late in 1766 was revealing, the American Society for Promoting and Propagating Useful Knowledge, held at Philadelphia. When the first permanent officers were elected in 1768, it became clear that leadership remained in the hands of the liberal Quakers and the near-Quakers despite the addition of numerous members who were neither. The society did itself honor when it elected Benjamin Franklin as its president but it gained little more, for Philadelphia's greatest scientist remained resident in England. More important was the vice-presidency, which went to Samuel Powel. Powel had been born and bred a Quaker and then, like so many young men of wealth and position, he had drifted into the Anglican communion. He, as a matter of fact, had been disowned by his meeting only a few months before his election. Charles Thomson, too, was given office in recognition of his vital contributions to the development of the society. Two other officers were Thomas Mifflin and Clement Biddle, who were Quakers at the time of their election but each of whom had to be disowned at the time of the Revolution because of his military activities. Isaac Bartram, another officer, remained a Friend throughout his life, although his brother Moses, also an active member of the American Society, joined the Free Quakers. Another

[6] Lewis R. Harley, *The Life of Charles Thomson* (Philadelphia [1900]), 202–203.

[7] William S. Middleton, "Philip Syng Physick, Father of American Surgery," *Annals of Medical History*, 1929, new ser., *1*:562–82.

[8] American Society Minutes, 23 May 1766, p. 2; William W. Hinshaw, *Encyclopedia of American Quaker Genealogy* (Ann Arbor, 1938), II, 464, 696; Charles Wetherill, *History of the Religious Society of Friends Called by Some the Free Quakers* (n.p., 1894), 11, 61; [Thomas Gilpin], *Exiles in Virginia* (Philadelphia, 1848), 67, 113.

office went to Dr. John Morgan, who was not a Quaker and who demonstrated little Quaker influence, although his mother had been a member of the Society of Friends until her marriage. Only one active officer seemed altogether untouched by the Quaker influence. He was Lewis Nicola, who had recently come to America from France by way of Ireland.[9]

The characteristics of the other society which united with the American Society in 1769 were quite different — although at the same time they did exhibit points of similarity. The second society asserted lineage from an abortive organization attempted in 1743, known then and upon its revival in 1767 as the American Philosophical Society. Initially, it had a strong Quaker cast. Its leaders in 1743 were John Bartram, the Quaker seedsman and botanist, Benjamin Franklin, and Dr. Thomas Bond, who had been disowned by his Quaker meeting only the year before. It was not until 1757 that John Bartram was disowned and even then he remained a Quaker at heart, continuing to attend meeting and continuing to use Friendly ways. Bartram, then, was a liberal Quaker, Bond a former Quaker in whom the Quaker influence was very strong, and Franklin a man with no Quaker background although with many personal, political, and business friends who were Quakers. Of the remaining six resident members, four were Quakers, only one of whom was later disowned.[10]

When in 1767, a new American Philosophical Society was established upon the foundations of the old, its membership displayed very different characteristics. It fell quickly into the camp of the Proprietary party, electing a former governor of the province as its president and numbering many Anglicans and Presbyterians among its members and particularly among its officers. Dr. Thomas Bond, to whom the society was a favorite "hobby horse," saw that it was impossible to curb the rising partisanship when one of his closest Quaker friends refused to accept membership because of his distaste for an Anglican priest and a Presbyterian minister who were given office.[11] The organization never lost its Proprietary-Anglican-Presbyterian orientation despite the acceptance of membership by many other Quakers. Nevertheless, Thomas Bond, the former Quaker, remained its guiding spirit throughout its short independent existence.

The 1769 union of the American Philosophical Society with the American Society to produce the American Philosophical Society, held at Philadelphia, for Promoting Useful Knowledge required careful balancing in the selection of officers. Where possible, previous officers of the American Society were balanced against previous officers of the American Philosophical Society. Because of his scientific eminence, Franklin was given the presidency — still in

[9] American Society Minutes, 13 Dec. 1766, p. 31; *ibid.*, 4 Nov. 1768, p. 132; Charles Thomson to Benjamin Franklin, 6 Nov. 1768, *New-York Historical Society Collections*, 1878, *11*:19; Hinshaw, *Encyclopedia*, II, 464, 460, 597.

[10] Benjamin Franklin to Cadwallader Colden, 5 April 1744, Albert H. Smyth, ed., *Writings of Benjamin Franklin* (New York, 1905–07), II, 277; John Bartram to Cadwallader Colden, 4 Oct. 1745, *New-York Historical Society Collections*, 1919, *52*:34; Mrs. Edith Verlenden Paschall very kindly made a transcript of the entries pertaining to the disownment of John Bartram in the Minutes of the Darby Monthly Meeting for the author's use.

[11] Dr. Cadwalader Evans to Benjamin Franklin, 27 Nov. 1769, Franklin Papers, *2*:201, American Philosophical Society.

absentia. It was the three vice-presidents who were chosen to serve as bridges between the two factions which comprised the society. Of these, Dr. Thomas Bond was the most effective. Well thought of by everyone, the former Quaker presided at most of the meetings. The Quaker influence was further in the background of Dr. Thomas Cadwalader, but it still lingered beneath the surface of that elegant, Anglican physician. Joseph Galloway, the third vice-president, was the leader of the Quaker party in the Assembly and very close to the Friends, though not of them. These men, once again, represented a leadership in the promotion of science in which the Quaker background was apparent.[12]

Men whose Quakerism was in the foreground of their life also played an important role in the success of the 1769 society — but a different sort of role. When the Quaker merchants joined the society in numbers, they furnished an economic and political support which, in very large measure, spelled the difference between the failure of the 1743 society and the success of the 1769 society. The Reverend Jacob Duché, an Anglican priest, remarked, "I was particularly pleased to hear from my friend, who is himself a Fellow of the Philosophical Society, that the Quakers had stepped forth and joined the votaries of Science; for their well known industry and application cannot fail, in all human probability, of ensuring its success." All denominations, he declared, "candidly confess, that no institutions have been carried on with so much spirit and crowned with so much success as those in which the Quakers have had the lead and direction." [13]

Perhaps, at this point, it would be pertinent to ask why it is necessary to express surprise at the discovery of considerable Quaker influence in the genealogy of America's first learned society, which — after all — was founded in the Quaker city. The fact is that the Philadelphia of 1769 is not well described as a Quaker city or Pennsylvania as a Quaker province. The Quakers were even then a minority — indeed, an inconsiderable minority. Burial records for 1769 indicate that the Quakers represented no more than 13 per cent of the city's population, and other facts suggest that they numbered even less than that.[14] In the province as a whole, the Reverend William Smith estimated that in 1759 the Quakers constituted only 20 per cent of the population.[15] Even those descended from Quaker families must have been relatively few in number, for the Quakers had become a minority in Pennsylvania as early as 1700.

Numbers, then, could not alone account for their prominence in the promotion of science; wealth was another matter. Patronage depended upon the concentration of wealth, and the Quakers had by 1769 succeeded in this pursuit better than any other group. Of the five richest men in the city, three were Quakers, and the fourth, Samuel Powel, had just turned Anglican.[16] The

[12] American Philosophical Society Minutes, 9 Feb. 1768, [*1*]:11, American Philosophical Society.

[13] Jacob Duché, *Observations on a Variety of Subjects, Literary, Moral, and Religious* (Philadelphia, 1774), 57.

[14] Robert Proud, *The History of Pennsylvania in North America* (Philadelphia, 1797–98), II, 340n.

[15] Dr. William Smith to Archbishop Secker, 27 Nov. 1759, E. B. O'Callaghan, ed., *Documents Relative to the Colonial History of the State of New York* (Albany, 1856), VII, 407.

[16] The 1769 tax lists are found in *Pennsylvania Archives* (3rd ser., Harrisburg, 1897), XIV, 886, 898, 952, 974, 1031. This source and

willing application of this wealth to the advance of science provided essential support.

In leadership, however, and in creative scientific work, the well-to-do Quaker did not excel. John Bartram — who was in a position to know — felt very strongly the failure of wealthy Philadelphians of all faiths to turn their interest to science. He remarked, "I sometimes observe that the major part of our inhabitants may be ranked in three Classes, the first Class are those whose thought and study is intirely bent upon geting and laying up large estates and any other attainment that dont turn immediately upon that hinge thay think is not worth thair notice. the second Class are those that are for spending in Luxury all thay can come at and are often the children of avaritious Parents, the third class are those that necessity obliges to hard labour and Cares for a moderat and happy maintainance of thair family, and these are many times the most curious tho deprived mostly of time and materials to pursue thair natural inclinations." [17] Bartram, Bond, Thomson, and most of their close associates would have fallen into the third class. Wealth was, at best, only a partial answer to the prominence of Quakers and near-Quakers in science.

The principal answer seems clearly to lie in the internal nature of the Quaker way. It lies in the positive attitude toward the study of nature, in the rational and empirical approach to thought, and it also lies in certain prohibitions which were shared. Jacob Duché asserted that Quaker disapproval of "all fashionable amusements and diversions, gives them liesure and opportunity of embarking in and prosecuting such schemes as are useful, as well as ornamental to human society." [18] John Bartram felt the force of this attitude when he declared that the 1743 American Philosophical Society would easily succeed if "we could but exchange the time that is spent in the Club, Chess and Coffee House for the Curious amusements of natural observations." [19] Men nurtured with such thoughts did not lose sight of them when they stopped attending the Friends meeting. Indeed, it is not impossible that men who gave up the meeting house were even more careful that they not waste their lives.

The American Philosophical Society of 1769 became the most important scientific body in the colonies, but it was not only in the organization of science that the Quaker influence was felt. It was similarly apparent in the background of men who made creative contributions to the advance of science. At the head of any list of Philadelphia scientists stands Benjamin Franklin, who was neither a Quaker nor a near-Quaker, despite the fact that he sometimes passed for a Quaker. He was too big to be so bracketed. After Franklin, the most important American "electrician" was Ebenezer Kinnersley, who was not a Quaker but a Baptist clergyman. Yet in three other men whose scientific attainments were popularly considered at least equal to those of Kinnersley,

---

others were used with similar results in Frederick B. Tolles, "Benjamin Franklin's Business Mentors: The Philadelphia Quaker Merchants," *William and Mary Quarterly*, 1947, 3rd ser., *4*:61.

[17] John Bartram to [Cadwallader Colden], 7 April 1745, Boston Public Library.

[18] Duché, *Observations*, 52.

[19] Bartram to Colden, 4 Oct. 1745, *N.-Y. Hist. Soc. Colls.*, 1919, *52*:160.

the Quaker influence was significant. They were James Logan, John Bartram, and David Rittenhouse.

James Logan made contributions to both the physical and the biological sciences. He published papers in the *Philosophical Transactions* and he published several separate pamphlets in Europe. He corresponded with leading European scientists. Recognition came to him particularly for his work on the pollination of Indian corn and his writings on optics. Logan was a Quaker, and in a sense, a Quaker of the Quakers. He was the most trusted advisor and confidant of the proprietor: secretary to William Penn, president of the council, acting governor, chief justice, he served Pennsylvania in many capacities and along the road managed to accumulate a fortune through his mercantile activities. On the other hand, his willingness to support and even to write in favor of defensive war put him in opposition to one of the most cherished principles of the society. Never disowned for this position, he was nevertheless marked as an independent and worldly Quaker. It was those who followed some of his arguments who were, years later, disowned for participating in the American Revolution. Logan was a liberal Quaker when it came to certain tenets of the Society.[20]

John Bartram received recognition from all parts of Western Europe for his accomplishments in finding and making known new forms of life in America — principally plant life. He was basically a taxonomist — even a field worker who left the naming of new specimens to others — but in a day when taxonomy was the reigning interest, his work was of the greatest importance. No other American did so much to extend the bounds of knowledge in this area. Bartram, it has been stated, was a Quaker in manner and in most of his thoughts even after his disownment. He, like Logan, was an independent thinker in religious matters who could not bring himself to accept the whole pattern of belief. In fact, Bartram stumbled over one of the most essential elements of the Christian faith — the divinity of Christ.[21]

David Rittenhouse was not a Quaker. He never had been. Yet, the Quaker mark was discernible in him. His mother was a Quaker and his father a Mennonite. Both of his wives were Quakers. In religious thought, he himself inclined more to the intellectual outlook of such dissenters as the Reverend Richard Price, but in his attitudes toward what was valuable in life, he indicated that he had drunk deeply of the Quaker philosophy. In the physical sciences, he was the most active figure in the Philadelphia area. Thereabouts, he gained a great reputation based largely upon his observations of the transit of Venus in 1769 and upon the accurate orrery or mechanical planetarium that he constructed.[22]

There are, then, some suggestions in the record as to why the Quaker influence was beneficial for the encouragement of science and for its advance.

[20] Wilson Armistead, *Memoirs of James Logan* (London, 1851) is unsatisfactory both as a collection and as a biography. The biography now in preparation by Frederick B. Tolles will supply a great need.

[21] Minutes of the Darby Monthly Meeting, 5 June 1758, 5 July 1758.

[22] The degree of accuracy attained by Rittenhouse in his orreries has for the first time been made clear in Howard C. Rice, Jr., *The Rittenhouse Orrery* (Princeton, 1954), 68–75.

There is even some indication of why it was more effective at the periphery of the Quaker circle than at its center. The answer, however, is not clear. The one certain fact is that there was a group of liberal Quakers, former Quakers, and near-Quakers who bore a large portion of the responsibility for the encouragement and advance of science in colonial Philadelphia.

Some Aspects of American Astronomy 1750–1815

By John C. Greene *

THE purpose of the present essay is to describe some of the main directions of observation and inquiry in American astronomy in the late eighteenth and early nineteenth centuries and to show how these researches reflected or modified prevailing conceptions of nature. Certainly no science exerted a profounder influence on Western thought in this period than astronomy. What atomic physics is to our own century Newtonian astronomy was to the eighteenth. In Europe the best scientific minds vied with each other to extend and perfect Newton's mathematical demonstration of the solar system. To obtain the necessary data, expeditions were sent to the far corners of the earth — to Lapland, to the Cape of Good Hope, to South America and the South Seas.[1] These combined efforts reached a climax with the publication, in the years 1798–1825, of Laplace's *Mécanique Célèste*, the summation of a century of progress in the astronomy of the solar system.

To these brilliant achievements American astronomers contributed no great discoveries either empirical or theoretical, but they kept abreast of the latest developments, made and published useful observations, and propounded theories of their own to account for what they observed. Astronomy was well established in the colonial colleges by the middle of the eighteenth century, especially at Harvard, where Professor John Winthrop continued with great ability the tradition of astronomical studies established at that institution in the seventeenth century. Appointed to the Hollis Professorship of Mathematics and Natural Philosophy in 1738, Winthrop sent to England for a copy of Newton's *Principia*, introduced the study of fluxions, expanded the use of experiments in instruction, and began regular observation of the heavens with the college telescope. His most important observations were published in the

* University of Wisconsin.

[1] Professor John Winthrop described to his class at Harvard some of the expeditions sent out to observe the transit of Venus in 1761: "The most Northern place the Transit was observ'd at, was in EUROPE, namely, Tornea in Lapland; almost under the polar circle. — In ASIA, it was observed at Tobolsk, the capital of Siberia, by a French Astronomer, who performed a journey thither of 4000 miles from Paris, at the instance of the Imperial Academy of Sciences at Petersburg, and under the auspices of the Czarina. . . . It was observed besides at Madras, which was farthest South-east, under the direction of the East-India Company of London. The French King also commissioned two members of his Royal Academy of Sciences, to make the observation in the East Indies. — In AFRICA, it was observed only at the Cape of Good Hope. It would have been so at St. Helena too, had not clouds prevented, by Astronomers sent to those places by the Royal Society, at the expence of his late Majesty K. George II. . . . In AMERICA, it was observed only at St. John's Newfoundland, and that at the expence of the Province of Massachusetts-Bay. And this place was the farthest West." John Winthrop, *Two Lectures on the Parallax and Distance of the Sun, as Deducible from the Transit of Venus* (Boston, 1769), 39–40. For a brief account of developments in astronomy in Europe in the eighteenth century see Peter Doig, *A Concise History of Astronomy* (London, 1950), chs. 8–9.

Philosophical Transactions of the Royal Society, of which he was elected a Fellow in 1766.[2]

The other colonial colleges did their best to emulate Harvard's example. Yale acquired an able mathematician and college president in Thomas Clap two years after Winthrop's appointment at Harvard. William and Mary made her bid in 1758 with the appointment of William Small to the post in natural philosophy. The College of Philadelphia had an astronomer of considerable talent in its first provost, the Reverend William Smith. After the Revolution it was not uncommon for college presidents to teach or study astronomy. Yale's Ezra Stiles was a lifelong student of the subject and a practicing observer. President Willard of Harvard corresponded with the Astronomer Royal in Britain and contributed several astronomical papers to the *Memoirs* of the American Academy of Arts and Sciences, which he had helped to found. On his death in 1804 he was succeeded by the professor of mathematics and natural philosophy, Samuel Webber. Robert Patterson occupied the mathematical chair at the University of Pennsylvania, but the Provost, the Reverend John Ewing, was equally capable as a natural philosopher, having served with Rittenhouse in surveying the boundaries of Pennsylvania. In the survey of the southern boundary, involving the extension of the Mason-Dixon line, they were joined by another college president versed in natural philosophy, the Reverend James Madison of William and Mary, one of the commissioners for the state of Virginia.[3]

Not all American astronomers were academics, however. The two best known in the early republic, David Rittenhouse and Nathaniel Bowditch, were both self-educated. Bowditch pursued his studies aboard ship while voyaging the seas as a common sailor. He later combined his nautical and astronomical knowledge to produce the famous *Practical Navigator*. Then, settled at Salem as president of a fire and marine insurance company, he devoted his leisure hours to preparing a translation with commentary of Laplace's *Mécanique Célèste*. Rittenhouse earned the means to study mathematics and astronomy by making precision clocks. His ingenious orrery, or mechanical model of the solar system, fetched a handsome price, and his growing reputation in both practical and theoretical astronomy procured him work as a surveyor. Andrew

[2] John Winthrop and several other figures discussed here are treated more fully in the writings of Frederick E. Brasch: "Newton's First Critical Disciple in the American Colonies — John Winthrop," in *Sir Isaac Newton, 1727–1927, a Bicentenary Evaluation of His Work* (Baltimore, 1928), 301–338; "The Newtonian Epoch in the American Colonies (1680–1783)," *Proc. Amer. Antiq. Soc.*, 1949, *49*: 314–332; "The Royal Society of London and Its Influence upon Scientific Thought in the American Colonies," *The Scientific Monthly*, 1931, *33*: 336–355, 448–469; "John Winthrop," *Publ. Astron. Soc. Pacific*, 1916, *28*: 153–170. Professor Samuel Eliot Morison discusses "The Harvard School of Astronomy in the Seventeenth Century" in *The New England Quarterly*, 1934, *7*: 3–24. See also Samuel A. Mitchell, "Astronomy During the Early Years of the American Philosophical Society," *Proc. Amer. Philos. Soc.*, 1943, *86*: 13–21; I. Bernard Cohen, *Some Early Tools of American Science . . .* (Cambridge, Mass., 1950), chs. 1–3; Solon I. Bailey, *The History and Work of Harvard Observatory 1839 to 1927 . . .* (New York and London, 1931), ch. 1; Theodore Hornberger, *Scientific Thought in the American Colleges, 1638–1800* (Austin, Texas, 1945), ch. 5.

[3] On the early history of science at Yale, see Louis W. McKeehan, *Yale Science The First Hundred Years* (New York, 1947). Horace W. Smith, *Life and Correspondence of the Rev. William Smith . . .* (2 vols., Philadelphia, 1880) contains scattered materials on the astronomical interests and activities of the first provost of the College of Philadelphia. See also the works cited in the note above.

Ellicott lived chiefly by surveying. He assisted Rittenhouse in running the southern, western, and northern boundaries of Pennsylvania; then, in 1791, he surveyed the tract for the new capital on the Potomac, aided by the Negro astronomer and almanac-maker, Benjamin Banneker. His most arduous survey was performed in the years 1797–1800, as commissioner to run the southern boundary of the United States according to the terms of the Pinckney-Godoy Treaty of 1796. One of the Spanish commissioners in the early stages of this survey was William Dunbar, gentleman planter of the Mississippi Territory and a scientist of no mean ability. But Dunbar returned to his estate near Natchez when the western end of the line had been fixed, and Ellicott was left to carry the line eastward through swamp and forest to the Atlantic coast. "It is to be presumed," wrote Ellicott in his account of this survey, "that no apology will be necessary, for any small inaccuracies which may be discovered in the astronomical observations, when it is considered that they were made at temporary stations, and the apparatus frequently exposed to the weather, for want of tents, and other covering; and almost as frequently so injured by the transportation from one place, to another, through the wilderness, that if I had not been in the habit of constructing, and making instruments for my own use, our business must have several times suspended, till the repairs could have been made in Europe." [4]

Until just before the Revolution, American astronomers had to depend on European journals for publication of their findings. In 1771, however, the American Philosophical Society began publication of its *Transactions.* The New Englanders, not to be outdone by Philadelphia, organized the American Academy of Arts and Sciences, which brought forth its first *Memoirs* in 1785. Observations of the transits of Mercury and Venus, of solar and lunar eclipses, of comets and meteors, as well as routine observations in the course of boundary surveys, occupied a prominent place in both publications. To these researches of American astronomers and to the speculations and reflections which they evoked we now turn.

1. Transits and Eclipses

Many of the astronomical papers published by Americans in these early years reported observations of transits and eclipses. The transits of Venus were of especial interest for reasons which Professor Winthrop explained to his students at Harvard in 1769:

> A TRANSIT OF VENUS UNDER THE SUN is the most uncommon, and the most important phaenomenon, that the whole compass of astronomy affords

[4] Andrew Ellicott, *The Journal of Andrew Ellicott, Late Commissioner on Behalf of the United States . . . for Determining the Boundary Between the United States and the Possessions of His Catholic Majesty in America . . .* (Philadelphia, 1814), 151. For biographical material see Catherine Mathews, *Andrew Ellicott, His Life and Letters* (New York, 1908); Edward Ford, *David Rittenhouse, Astronomer-Patriot 1732–1796* (Philadelphia, 1946); Robert E. Berry, *Yankee Stargazer, The Life of Nathaniel Bowditch* (New York and London, 1941); Mrs. Dunbar Rowland, comp., *Life, Letters and Papers of William Dunbar of Elgin, Morayshire, Scotland, and Natchez, Mississippi, Pioneer Scientist of the Southern United States* (Jackson, Mississippi, 1930); Shirley Graham, *Your Most Humble Servant* [Benjamin Banneker] (New York, 1949).

us. So uncommon is it, that it can never happen above twice in any century; in others, but once; and in some centuries it cannot happen at all. And the importance of it is such, as to supply us with a certain and complete solution of a very curious Problem, which is inaccessible any other way. . . .

It is plain enough, that our hopes of finding the distances of the heavenly bodies, with any certainty, must be built on observations of their parallaxes. If the diameter of the Earth bear any sensible proportion to the distance of an heavenly body, that body must be subject to a parallax, of some quantity or other; that is, it must appear in different points of the starry heaven, when view'd from different parts of the Earth. . . . Venus in her inferior conjunction is but little more than one quarter so far from us as the Sun is, and therefore her parallax almost four times as great as his. This therefore is the most advantageous circumstance. But a Transit of Venus is the most favorable conjuncture of all, because the limbs of the Sun afford the best terms with which to compare the planet; and instead of trying to observe the parallactic angles, which are extremely small, it is much better to observe the differences of time, occasion'd by them, which are much more sensible. . . . The best observations will be, when the planet is in contact with the Sun's limbs, at its immersion and emersion; the moments of which may be determined with great accuracy, if the air be clear, by such as are furnished with good astronomical instruments, and are expert in the use of them.[5]

Winthrop could speak with authority on this subject, for he was an old hand at observing transits. Two of his earliest communications to the Royal Society concerned transits of Mercury, and in 1761 he had organized and carried through an expedition to Newfoundland for the purpose of observing the transit of Venus. The governor and legislature of Massachusetts Bay were apprised of the fact that astronomers had been looking forward to this event with great anticipation and that royal support had been secured in England and France for expeditions to the remote corners of the earth to observe the phenomenon. They were told further that Newfoundland was the only British possession in North America from which the transit could be viewed and that the expenditure of public funds for this purpose would advance the interests of commerce and navigation as well as those of science. Duly impressed, the legislature authorized the use of the province sloop to transport the expedition to St. John's, Newfoundland.

Winthrop and his student assistants arrived at St. John's on May 19th, established a tent camp on a hill overlooking the city, and proceeded to make the necessary preliminary arrangements and observations. The morning of June 6th dawned clear and calm, and they had the "high satisfaction" of seeing Venus on the sun and of showing it to the cluster of local gentry who had gathered for the occasion. Delighted with the sight, these gentlemen resolved to name the hill on which they were standing Venus's Hill in honor of the event. Winthrop's observations appeared in due course in the *Philosophical Transactions* of the Royal Society. The comparison of these data with those recorded in other parts of the world would, said Winthrop in his report of the expedition, "give the true path of Venus, abstracted from parallax; by which means, the quantity of the parallax will at length be discovered. The right

[5] Winthrop, *Two Lectures on the Parallax and Distance of the Sun*, 5, 18, 20–21.

determination of which point will render this year 1761 an ever memorable year in the annals of astronomy." [6]

The transit of Venus in 1769 produced an intercolonial scientific effort of major proportions. The American Philosophical Society, with help from the College of Philadelphia, the Pennsylvania Assembly and the Penn family, stationed observers at State-House Square in Philadelphia, at Rittenhouse's house in Norriton, and at Cape Henlopen, Delaware. In Rhode Island, the transit was observed by Ezra Stiles in Newport and by Benjamin West and Joseph Brown in Providence. At Harvard, Winthrop was ready with the fine new instruments recently procured from England through Franklin's good offices. Samuel Williams, who had accompanied Winthrop to Newfoundland and was to succeed him in the Hollis professorship, availed himself of the telescope of Tristram Dalton, Esq., at his country house in West Newbury, Massachusetts. In Baskenridge, New Jersey, the event was recorded by William Alexander, who styled himself Lord Stirling. In Canada there were several observers. Winthrop sent his observations to the Royal Society as usual, but West and Williams sent theirs to the American Philosophical Society for publication with those of Rittenhouse and his colleagues and the observations communicated from the Royal Observatory at Greenwich. From these and other data Rittenhouse and William Smith calculated the solar parallax and found it to agree substantially with that obtained from the best observations of the transit of Venus in 1761.[7]

The total solar eclipse of June 16, 1806, was observed by a score of American astronomers scattered from Brunswick, Maine, to Natchez, Mississippi.

[6] John Winthrop, *A Relation of a Voyage from Boston to Newfoundland, for the Observation of the Transit of Venus, June 6, 1761* (Boston, 1761), 21 (microfilm). If the solar parallax could be accurately determined, said Winthrop, "the distance of the Sun, and of all the Planets, and of all the Comets would be known too; and their magnitudes would also be known, from their apparent diameters. This would give us a just idea of the vast dimensions of the solar system, and of the mighty globes which compose it. Nor can we, but by such observations, know whether the Earth continues to revolve at the same distance from the Sun, or whether it gradually approaches him, as there is some reason to suspect; nor whether the Sun remains of the same magnitude, or consumes away and is diminish'd by the light which he is incessantly sending forth." *Ibid.*, 5.

[7] *Trans. Amer. Philos. Soc.*, 2nd ed., 1789, *1*: 4–126; *Philos. Trans. Roy. Soc. London*, 1769, *59*: 289–330, 351–358; *Mem. Amer. Acad. Arts Sci.*, 1815, *3*: part 2, 288–292. Benjamin West also printed his observations separately in a pamphlet entitled: *An Account of the Observation of Venus upon the Sun, the Third Day of June, 1769, at Providence, New England . . .* (Providence, R. I., 1769). An earlier effort to arrange for multiple observation of a transit took place in 1753, when Benjamin Franklin printed and sent off to various places in British America fifty copies of some "Letters Relating to a Transit of Mercury over the Sun, Which Is to Happen May 6th, 1753." These letters, containing instructions for observing the transit, were originally drawn up by the French astronomer de Lisle and dispatched to Quebec via New York. In New York they were seen by James Alexander, who translated them and sent copies to Franklin in Philadelphia. Cloudy weather defeated the attempts of Franklin and Alexander to observe the transit, but in Antigua one William Shervington, having received the letters of instruction from Franklin's nephew there, sent in observations of the event which were printed in the *Philosophical Transactions* of the Royal Society of London for the year 1753. See Jared Sparks, ed., *The Works of Benjamin Franklin . . .* (Chicago, 1882), *6*: 159–161, 187–188. I. B. Cohen, "Benjamin Franklin and the Transit of Mercury in 1753 . . . ," *Proc. Amer. Philos. Soc.*, 1950, *94*: 222–232, contains a facsimile of Franklin's communication and a full discussion of the circumstances and results connected with its issuance. James Alexander, a prominent lawyer and government official in New York, was the father of William Alexander, mentioned as an observer in New Jersey. William assumed the title "Lord Stirling," claiming descent from Sir William Alexander, favorite of James I. Failing to win recognition of his claim in the House of Lords, he continued to style himself "Lord Stirling" and was so known during his able career as one of Washington's generals during the Revolution.

Bowditch observed it in his garden at Salem, Samuel Williams at Rutland, Vermont, Professor Parker Cleaveland at Bowdoin College; other reports reached Bowditch from Falmouth, Martha's Vineyard, and Nantucket. Simeon De Witt observed the eclipse at Albany and sent a painting of it, done by a local portrait painter, to the American Philosophical Society. At Kinderhook, New York, the Spanish astronomer, José Joaquin de Ferrer, was an interested observer. Patterson and Ellicott stationed themselves at Philadelphia and Lancaster respectively, and Ellicott received from Natchez the observations of his former surveying companion, William Dunbar. These various findings, some of them sent to Boston and some to Philadelphia, were brought together by Bowditch in the *Memoirs* of the American Academy in 1809. In subsequent volumes Bowditch assembled the scattered observations of other transits and eclipses and sought to determine their elements.[8]

The effect of these multiplied observations and calculations of transits and eclipses was to confirm the growing conviction of the regularity and wise contrivance of nature. The determination of the sun's parallax, Winthrop assured his Harvard class, would result in "a deeper insight into many of the wonderful works of GOD," illuminating moral as well as natural philosophy. Nearly a half century later the prediction and observation of the total eclipse of 16 June 1806 was the occasion for similar comment by a writer in the *Panoplist*, a Congregationalist magazine. There was little likelihood, this writer declared, that anyone who witnessed the late eclipse would be disposed to deny the existence of God. The danger was rather that eclipses, "because they are perfectly agreeable to the regular course of nature, and can be demonstrated to result from established laws," might cease to excite pious admiration.[9] The prediction of celestial events was losing its novelty.

2. Comets

Comets had claimed their share of scientific interest in America from the days of Samuel Danforth's *Astronomical Description of the Late Comet or Blazing Star*, published in 1665. The comet of 1759 was deemed of sufficient importance at Harvard to occasion two special lectures to the student body by Professor Winthrop. He began, as Danforth had in the previous century, by dispelling erroneous notions concerning comets, declaring "that they are neither below the moon, nor among the fix'd stars; that they consist not of any kind of exhalations; — that their tails are not produced by the transmission of light thro' their bodies; nor by any refraction of light, caused either by their atmospheres or by the celestial matter through which it passes; — that the tails are not turned opposite to the Sun by the impulse of his rays; and lastly, that the motion of Comets is not in strait lines." [10] He then proceeded to ex-

[8] *Mem. Amer. Acad. Arts Sci.*, 1809, *3*: part 1, 18–32; 1815, part 2, 275–279; *Trans. Amer. Philos. Soc.*, 1809, *6*: part 2, 255–277; 293–303. José Joaquin de Ferrer was a Spanish naval officer who spent considerable time in the West Indies and North America making astronomical observations for geographical purposes. He published several papers in the *Transactions* of the American Philosophical Society.

[9] "Serious Thoughts Excited by the Late Eclipse," *The Panoplist and Missionary Herald*, 1807, *2*: 84.

[10] John Winthrop, *Two Lectures on Comets, Read in the Chapel of Harvard College . . . in*

pound the Newtonian theory of comets. The present comet, he declared, was predicted on Newtonian principles by Edmond Halley, who foretold the reappearance late in 1758 or early in 1759 of the comet observed by Kepler in 1607 and by Halley himself in 1682.

> . . . This prediction we have now the satisfaction to see verified; the present Comet agreeing hitherto so well with the former, as to leave little room for doubt that they are the same. We may therefore, upon sure grounds, expect that the rest of the comets will return too; tho' we are not able as yet to prefix the years of their returns. . . A theory, which so accurately answers to the motions of all the Comets which have been accurately observed, even to some which have been very extraordinary, and which accounts for them by the same laws as the motions of the Planets are accounted for, cannot but be true.[11]

Having expounded and defended the Newtonian theory of comets, Winthrop turned to the subject of their natural and moral uses. Morally, he declared, they serve to remind men of the Supreme Governor of the Universe and of his wisdom, power, and benevolence. Just as the uniform direction of rotation and revolution among the planets and their satellites evinces intelligent design, so the eccentric orbits of the comets and the varying inclinations of their planes are evidently contrived to prevent collisions and undue gravitational perturbations. It is unlikely, he continued, that the comets are inhabited, in view of the tremendous alternations of heat and cold, light and darkness, to which they are subject as they approach and leave the sun. But perhaps, as Newton suggested, the tails of comets replenish the atmospheres of the planets. Possibly, too, the comets themselves eventually fall into the sun and thus prevent the diminution of its mass and fuel supply.

Winthrop was too much a Puritan, however, to leave the subject of comets without considering their possible use as agents of divine justice. William Whiston, he declared, had produced an impressive array of evidence in support of his cometic theory of the Deluge. Whether that theory were accepted or not, there could be no doubt that the near approach of a comet to the earth would produce dreadful consequences. That no such disaster had occurred, excepting possibly the Deluge, was itself a testimony to the wisdom of the omnipotent Creator. Divine Providence, Winthrop concluded, would preserve the frame of nature undisturbed so long as it served the purposes for which it was created. "Longer than that, it is not fit that it should subsist." [12]

April 1759 on Occasion of the Comet Which Appear'd in That Month (Boston, 1759), 19. Compare Danforth in 1665: "If the Comet be no vapour but a *celestial planetick luminary*, moving constantly in its *Eccentrick orb*, and if the stream thereof be no real flame, but the *irradiation* of the *Sun* through the Comet's head, it will necessarily follow that the Comet is not *consumed*, *dissipated*, or *extinguished*, but rather ascended toward its *Apoge*, *i.e.*, the farthest point distant from the Earth, and so being buried in the *deep abyss* of the Heavens, becomes inconspicuous to us." *An Astronomical Description of the Late Comet or Blazing Star . . .* (Cambridge, Massachusetts, 1665), 15–16 (photostat copy). See Brasch, "Newton's First Critical Disciple," cited above, for an account of Winthrop's other writings on comets, especially his "Cogitata de Cometis," *Philos. Trans. Roy. Soc. London* 1767, 57: 132–154.

[11] Winthrop, *Two Lectures on Comets*, 29–30.

[12] *Ibid.*, 44. William Whiston was Newton's successor in the Lucasian professorship at Cambridge. In 1696 Whiston published *A New Theory of the Earth, from Its Original to the Consummation of All Things, Wherein the Creation of the World in Six Days, the Universal Deluge, and the General Conflagration, As Laid Down in the Holy Scriptures, Are Shewn to Be Perfectly Agreeable to Reason and Philosophy.* In this work he attempted to prove, among other things, that the Deluge was caused

Winthrop's exposition of the science of comets was undeviatingly Newtonian. Some Americans, however, were disposed to question Sir Isaac's views concerning the physical condition and intended uses of comets. To Dr. Hugh Williamson it seemed preposterous to suppose that comets could have been formed "for the sole purpose of being frozen and burnt in turns." His reflections on the subject, provoked by the comet of 1769, were communicated to the American Philosophical Society in November, 1770, and subsequently published in its *Transactions*.[13] Newton was wrong, said Williamson, in supposing that the temperature of planets and comets varies inversely with the square of their distances from the sun. The frigidity of mountain climates refutes this idea and shows the extent to which the heat-producing power of the sun is affected by the relative density of the atmosphere. But the atmosphere of a comet is more or less dense in proportion to its distance from the sun. When it approaches perihelion, its atmosphere is repelled and elongated by the sun's rays in such a manner as to insure the comfort of its inhabitants despite their proximity to the solar furnace. Then, as the comet recedes, its atmosphere gradually re-encompasses the nucleus in great depth and insures transmission of solar heat to the comet's surface at the farthest reaches of the orbit. These considerations, Williamson argued, show that comets must be inhabited, and analogy renders it likely that the denizens of those comets which travel farthest into space are vastly superior to man in longevity and in knowledge of God's works.

Williamson's essay found its way into a weekly newspaper, where it was seen by Andrew Oliver, Jr., son of the Secretary of the Massachusetts Bay Colony. Oliver was a mathematician and astronomer of considerable talent, having devoted what time he could spare from his duties as judge of the inferior court at Salem to scientific interests acquired under Winthrop's tutelage. Attracted by Williamson's notion that comets are habitable, he undertook to establish it on firmer ground by assigning an adequate physical cause for the atmospheric transformations by which the surface temperature of comets is adjusted to the needs of living beings. In his *Essay on Comets*, published at Salem in 1772, he advanced the hypothesis that the tail of a comet results from the mutual repellency between the comet's atmosphere and that of the sun. Assuming the force of repulsion to vary inversely as the quantities of the repellent fluid con-

---

by the near approach of a comet to the earth. He later identified the comet as Halley's comet. Winthrop was not alone in being well impressed with Whiston's speculation. Newton "well approved" of it, and Locke thought the author "more to be admired that he has laid down an Hypothesis whereby he has explained so many wonderful, and before inexplicable Things in the great Changes of this Globe, than that some of them should not easily go down with some Men; when the whole was intirely new to all." William Whiston, *Memoirs of the Life and Writings of Mr. William Whiston . . .* (London, 1749), 43–44.

[13] Hugh Williamson, "An Essay on the Use of Comets, and an Account of Their Luminous Appearance; Together with Some Conjectures concerning the Origin of Heat," *Trans. Amer. Philos. Soc.*, 2nd ed. (Philadelphia, 1789), *1*: 133–143. Williamson was born in Pennsylvania and studied medicine in Europe. Elected to the American Philosophical Society in 1768, he assisted Rittenhouse in observing the transits of Mercury and Venus in 1769. As a delegate from North Carolina he played an active role in the formation of the Federal Union. In 1793 he settled in New York and devoted himself to scientific and literary pursuits. It should be noted in passing that the notion of the habitability of comets was not at all uncommon. It accorded with the more general presupposition that all material bodies must perform some function subservient to the activities of intelligent beings. See Arthur O. Lovejoy, *The Great Chain of Being . . .* (Cambridge, Mass., 1950), ch. 4.

tained in the atmospheres, the atmosphere of the comet would be driven to a great distance by the sun's atmosphere as the comet made its approach. After perihelion, however, the repulsion would steadily decrease, and gravitational attraction would draw the comet's atmosphere tightly around the nucleus again and thus prepare it to sustain life at a great distance from the sun. In support of this theory, Oliver described some electrical experiments with an artificial comet — "a small, gilt cork ball, with a tail of leaf-gold, about two inches and an half in length." He was careful to explain, however, that the mutual repulsions of atmospheres, though analogous to the electrical repulsion of the balls in the experiment, was not itself a purely electrical phenomenon. If it were, there would be destructive electrical discharges between planets and the tails of comets — an unthinkable event, "unless we can suppose, that infinite wisdom and goodness would create one world, merely for the destruction of another." [14]

Throughout his essay Oliver took pains to allay the fears which had been aroused by the comet of 1769 and by speculation in the newspapers concerning the disasters which might attend it. He denied that a comet's tail could deluge the earth with water. The rarified atmosphere of the recent comet, he declared, could scarcely have provided enough water for a common thunder shower. Nor should comets be regarded as penal worlds, for it would be absurd to suppose that God created fifty or more such worlds to provide a place of punishment for the inhabitants of only six planets. How much more consistent with His divine attributes to suppose that He adapted comets as well as planets as residences for intelligent beings by the simple device of making the quantity and quality of their atmospheres vary with their distances from the sun. Indeed, said Oliver, if the repellent fluid presently contained in the atmospheres of the bodies of the solar system were diffused evenly in the space occupied by that system and those bodies were then introduced into that repellent medium, each body would by the laws of gravitation attract to itself the amount of fluid now contained in its atmosphere.

> . . . If the Comets be supposed to have been created and projected in their several orbits, at their aphelia, or at their greatest distances from the Sun, it may be easy upon this hypothesis to account for their having atmospheres so much exceeding those of the Planets in their dimensions, for providence has so ordered it, that the angles of the inclinations of their orbits to the eclyptic and to each other are generally very great, and their motions are directed to all parts of the Heavens indiscriminately, whereby their distances from the Planets and from each other at their aphelia, are great beyond human conception; consequently, they were at liberty to share amongst themselves, without any molestation from the Planets, all that part of the fluid, which filled the vast spaces of the System, without the planetary regions; therefore if the hypothesis be granted, they must necessarily have such atmospheres, as, in fact, we find they have, and which, in their descent through the planetary spheres are, by the (supposed) repulsions of the Sun's atmosphere, driven to such astonishing distances behind them, as occasion may require. Those whose aphelion distances were greatest, being more

[14] Andrew Oliver, *An Essay on Comets, in Two Parts* . . . (Salem, Mass., 1772), 47–48; Oliver's *Essay* was reprinted in 1811 under the same cover with Winthrop's *Two Lectures on Comets*. (John Davis, ed., *Two Lectures on Comets, by Professor Winthrop, Also an Essay on Comets, by A. Oliver jun. esq., with Sketches of the Lives of Prof. Winthrop and Mr. Oliver. Likewise a Supplement, Relative to the Present Comet of 1811* (Boston, 1811).

solitary would condense round them the greatest atmospheres, and such, their greater distances from the Sun would require, upon the foregoing principles, to make them comfortable habitations." [15]

Questions connected with comets continued to excite a certain amount of discussion in the years after the Revolution. The fifth edition of Morse's *American Universal Geography*, issued in 1805, cited Oliver's *Essay* with approval, declaring:

> Modern astronomy shows the terror and dismay, which comets once occasioned to have been groundless, and teaches us to view them, as well as other celestial appearances, with composure. While ignorant of nature, we are easily alarmed at her operations, but acquaintance with the constance and regularity with which she proceeds will tend to quiet our apprehension and inspire us with confidence; fears, which hovered in the darkness that covered her, fly before the rising light of science. Among all the comets hitherto observed, there is not one, which, according to the knowledge we have of it, will probably produce any fatal effects on our earth: and it may, perhaps, be allowable to conjecture, that farther discoveries in this part of astronomy will lessen the probability of danger, or increase that of safety.[16]

The editors of the *Medical Repository* were equally complacent with respect to the comet which appeared in 1807, declaring themselves content to await the reports of "our Dewitts, Pattersons, Ellicotts, Madisons, Garnetts, our Ferrers, our Webbers" to learn whether this particular comet was one of the four hundred and fifty already assigned to "our department of the universe." Accounts were soon available. Dunbar reported from Natchez, Ferrer from Havana, Folger from Nantucket, Professor Farrar from Cambridge, and Bowditch from Salem. From the New England observations Bowditch calculated the elements of the comet's orbit according to the method laid down by Laplace and pronounced the comet a new one.[17]

3. Meteors

On 14 December 1807, while the comet was still visible, an even more spectacular event occurred in the heavens. At half past six in the morning Judge

[15] *Ibid.*, 86–87.

[16] Jedidiah Morse, *The American Universal Geography* . . . 5th ed. (Boston, 1805), *1*: 32–33.

[17] *The Medical Repository*, 1808, *11*: 197–198; *Mem. Amer. Acad. Arts Sci.*, 1809, *3*: part 1: 1–17; *Trans. Amer. Philos. Soc.*, 1809, *6*: part 2: 345–347; 368–375. Col. Jared Mansfield, Surveyor-General of the United States, observed the comet in Cincinnati. His findings were published in the *Memoirs* of the Connecticut Academy of Arts and Sciences, 1810, *1*: part 1: 103–110. The comet of 1811 was observed throughout the United States with even greater attention than that of 1807. "Few Comets," said Professor Day of Yale, "have presented themselves to our view, under circumstances more favourable, for observing their motions." Concerning the various theories which had been advanced to explain the tails of comets, Day observed: "Some extravagance of conception is certainly excusable, in attempting to explain the constitution of a luminous object, which occupies a greater space, than all the other bodies in the solar system. But the schemes which have hitherto been proposed, for this purpose, are rather to be considered as displays of the power of imagination, than specimens of the exercise of sound and sober reason. Those who have a taste for these visionary hypotheses, may easily contrive them for themselves; or may find, in the common astronomical works, a very convenient assortment of them, adapted to the fancy, of almost every description of readers." *Mem. Conn. Acad. Arts Sci.*, 1813, *1*: part 3: 352. See also the *Memoirs* of the American Academy, 1815, *3*: part 2, 308–325, for the reports of Bowditch, Farrar, and others on the comet of 1811.

Nathan Wheeler, walking outside his house in Weston, Connecticut, sensed a flash of light overhead and, looking into the northern sky, perceived a fiery meteor rising swiftly from the horizon, burning with a vivid light. It paled as he watched it and disappeared before it reached the zenith. A half minute later three loud reports shook the air, followed by a volley of lesser reverberations. The meteor was seen as far south as New York and as far north as Berkshire county, Massachusetts, and the explosions were heard and felt fifty miles away. At each report fragments of the meteor were hurled to the ground, arousing the inhabitants to investigate the cause of the heavy thudding. By the time Professors Silliman and Kingsley arrived from Yale College, many of the fragments had been carried away. The largest, weighing about thirty-five pounds, had been hammered to pieces and heated in a crucible in the hope of extracting precious metal from it. Nevertheless, the Yale professors managed to collect enough specimens to make a satisfactory analysis of the meteor. A preliminary account was published in the *Connecticut Herald* to satisfy public curiosity; full scientific details were then forwarded to the American Philosophical Society for publication in their *Transactions* and, subsequently, in foreign scientific journals.[18]

The Silliman-Kingsley paper was by no means the first American contribution to the study of meteors. Professor Winthrop had submitted accounts of several meteors to the Royal Society, exchanging opinions with Dr. John Pringle as to their nature and causes. The most ambitious theoretical approach to the subject, however, was Thomas Clap's *Conjectures upon the Nature and Motion of Meteors, Which Are Above the Atmosphere*, published posthumously in 1781. From various accounts of "high meteors" and from the known laws of physics Clap reasoned that they must be about half a mile in diameter and solid, at least in their external parts, since they were able to withstand the shock of tremendous explosions without disintegrating or losing their globular shape. The rumbling noise heard during their passage he attributed to friction with the earth's atmosphere: "for a cannon ball, of six inches in diameter, passing through the air, with 1–25th part of the velocity of the Meteor, will make a humming noise, which is generally heard two miles." The same friction, he argued, must generate an electrical charge capable of producing the explosions reported by observers. No known laws of nature, however, could raise such bodies to a height of one hundred miles above the earth's surface and give them a projectile velocity twenty times that of a cannon ball. They must, therefore, like comets and planets, move through space in orbits determined by the laws of projectile and centripetal force, "and as all the coelestial bodies are so remote that they can have no sensible influence upon them, when they are within 100 miles of the earth, it is evident that the earth must be the attractive central body, round which they revolve, as the secondary planets revolve round

[18] Benjamin Silliman and James Kingsley, "Memoir on the Origin and Composition of the Meteoric Stones Which Fell from the Atmosphere . . . 14th December 1807," *Trans. Amer. Philos. Soc.*, 1809, *6*: part 2, 323–345. The account appeared in several newspapers and was eventually reprinted in the *Memoirs* of the Connecticut Academy.

the primary, or rather as comets revolve round the sun in long elipses, near to a parabola." [19]

Clap then proceeded to calculate the motion of meteors around the earth, adopting the method used by Halley for comets. "This calculation," he declared, "seems to answer exactly to all the apparent motions of these Terrestrial Comets, and particularly that they appear first to be 50 or 100 miles distant from the earth, then, in their course, to come within 20 or 30 miles of it, and afterwards are at a greater distance again." The number of such bodies, he conceded, could not be determined precisely with the available data, but it must fall within certain limits. "Let us . . . conjecture for the present, until we have farther light by more accurate observations, that there are 3 such Comets revolving round the earth, whose mean distances are about as great as the moon's, and, therefore, performing about 36 revolutions in a year; then one of them will appear in each country of 500 miles square, once in 27 years: And so often at least, they have been in fact seen in Old England and New.[20]

A difficulty suggested itself, however: what was to prevent the friction of the earth's atmosphere from progressively retarding the velocity of these meteors until they fell to earth? This eventuality, Clap suggested, might be prevented by the meteor's recoil from the earth when the electrical explosions took place, or possibly Providence had ordained that the moon or some other heavenly body might act to accelerate the motion and enlarge the orbits of meteors from time to time, just as, according to Whiston, the comet which caused the Deluge accelerated the motion of the earth and increased the period of its annual revolution five and a quarter days. Likewise, the explosions of meteors might serve to cleanse and purify the atmosphere of the earth, rendering it more salubrious for living beings.

It was to Clap's theory that Silliman and Kingsley turned for an explanation of the phenomena connected with the Weston meteor. After rejecting various rival theories attributing meteors to atmospheric combinations, the action of lightning on masses of common stone, or the explosions of volcanoes, they gave tentative approval to Clap's conception of meteors as terrestrial comets, acknowledging, however, that it was not free from difficulty. The arguments in support of this position were set forth at greater length by their colleague, Professor Jeremiah Day, in a paper on the origin of meteoric stones, published in the *Memoirs* of the Connecticut Academy of Arts and Sciences in 1810.[21] Meteors, said Day, could not be formed in the atmosphere or in terrestrial volcanoes, for in neither case would the size, chemical composition, velocity, or path of the projectile agree with that observed in actual meteors. Such volcanic or atmospheric productions would fall to earth, but there was nothing to suggest that meteors did so. The diameter of the Weston meteor, judging from reports of its apparent diameter, could not have been less than several hun-

[19] Thomas Clap, *Conjectures upon the Nature and Motion of Meteors, Which Are Above the Atmosphere* (Norwich, Connecticut, 1781), 11. This essay was found among Clap's papers after his death and printed by private subscription.

[20] *Ibid.*, 12.

[21] Jeremiah Day, "A View of the Theories Which Have Been Proposed, to Explain the Origin of Meteoric Stones," *Mem. Conn. Acad. Arts Sci.*, 1810, *1*: part 1, 163–174.

dred feet, hence the main body of the meteor could not have been seriously diminished or deflected from its course by the loss of the fragments found in Connecticut. It was possible, Day conceded, that a mass of stone projected from a volcano on the moon might be drawn into the earth's gravitational field and describe a path similar to that observed in meteors, but it was unlikely, to say the least, that thousands of such projectiles, each several hundred feet in diameter, were thrown off from the moon annually. Clap's theory, on the other hand, was consistent with observed phenomena. If it involved some unverified assumptions concerning electricity, these could be tested by experiment. Even so, it could not be regarded as anything more than a plausible hypothesis.

Bowditch arrived at much the same conclusion from his study of the Weston meteor. He collected accounts of it with great diligence, journeying to Wenham, Massachusetts, to record as accurately as possible the observations of one Mrs. Gardner, who had had an excellent view of the meteor from her bedroom window. By comparing these data with those supplied by Silliman and Kingsley and with others sent in from Rutland, Vermont, he was able to estimate the height, direction, velocity, and magnitude of the meteor. He placed the velocity at more than three miles per second and estimated the weight, supposing the meteor to be of the same specific gravity as the specimens described by Silliman and Kingsley, at more than six million tons. There could be no question, he declared, that the fragments which fell in Connecticut were but an inconsiderable fraction of the meteor, which must, therefore, have continued on its course without falling to earth. The true explanation of these phenomena he left an open question: "The greatness of the mass of the Weston meteor does not accord either with the supposition of its having been formed in our atmosphere, or projected from a volcano of the earth or moon; and the striking uniformity of all the masses that have fallen at different places and times (which indicates a common origin) does not, if we reason from the analogy of the planetary system, altogether agree with the supposition that such bodies are satellities of the earth." [22]

Not all Americans were ready to give up the volcanic theory of the origin of meteoric stones, however. If they came from outer space, reasoned Dr. John Brickell of Charleston, South Carolina, they would gradually increase the mass of the earth, thereby causing it to revolve closer to the sun with accelerated motion. Moreover, if the increase in the earth's mass took place at the expense of the moon or other bodies of the solar system, corresponding alterations would take place in their orbits. "None of these consequences having occurred," said Brickell, "we must infer, the quantity of matter in the earth is unchanged since the creation, and consequently, that these aeropiptick stones are thrown from our volcanoes." [23]

[22] Nathaniel Bowditch, "An Estimate of the Height, Direction, Velocity and Magnitude of the Meteor, That Exploded over Weston in Connecticut, December 14, 1807 . . . ," *Mem. Amer. Acad. Arts Sci.*, 1815, *3*: part 2, 236.

[23] Letter from Dr. John Brickell, Savannah, Georgia, to Josiah Meigs, 22 February 1809, quoted in *The Monthly Anthology and Boston Review*, 1809, *6*: 283. A somewhat different argument against the lunar origin of meteoric stones was advanced, with mathematical demonstrations, by Professor F. R. Hassler at West Point. After showing that the velocity of the projectile from the moon would have to be

4. The Stability of the Heavens

As Brickell's argument suggests, the general conviction of the regularity and wise contrivance of nature was disturbed from time to time by evidence or theoretical implications suggesting the mutability of the heavens. Halley had noted as early as 1718 that several stars, including Arcturus and Sirius, had shifted their positions in the heavens since Ptolemy catalogued them, and Tobias Mayer, in 1756, listed fifty-seven stars whose positions varied from those which Olaus Römer had assigned to them fifty years earlier. In 1750 Thomas Wright speculated that the Milky Way was a great system of stars moving about some central globe more or less in the same manner as the planets move around the sun. Five years later his suggestion was adopted by Immanuel Kant and expanded into a full-blown theory of stellar evolution. Late in the century Herschel began producing observational evidence to support his theory of the gradual formation of stars and star systems from nebulous material scattered through the heavens, and Laplace proposed his famous nebular hypothesis to account for the origin of the solar system.

The idea of the mutability of the heavens made slow headway, however. In American colleges the third edition of James Ferguson's *Astronomy Explained* (1764) was the standard text in the late eighteenth century. In it the prevailing view of the heavens was set forth with the usual rhapsodies: "Thousands of thousands of Suns, multiplied without end, and ranged all around us, at immense distances from each other, attended by ten thousand Worlds, all in rapid motion, yet calm, regular, and harmonious, invariably keeping paths prescribed them; and these Worlds peopled with myriads of intelligent beings, formed for endless progression in perfection and felicity." [24] The uniformities in the arrangement and motions of the planets were cited as proofs of divine contrivance, while such evidences of instability as the apparent shortening of the moon's periodical month were used to refute the doctrine of the eternity of the world. The fixed stars, Ferguson explained, were so called "because they have been generally observed to keep at the same distance from each other." Arcturus and some others were apparent exceptions, but it would require the observations of many ages to determine whether their motions were real. The disappearance of stars and the appearance of new ones could not be denied but could, perhaps, be explained satisfactorily:

> . . . It would seem that the periodical Stars have vast clusters of dark spots, and very slow rotations on their Axes; by which means, they must disappear when the side covered with spots is turned towards us. And as for those which

more than ten times the velocity of the moon in its orbit, Hassler asked: "Can we believe that there exists in the moon any internal power, capable of producing this effect? When we consider how small the attraction of gravitation is at the moon, would not the existence of such a projectile force prove in the lapse of ages, destructive to that body? And when centuries, and even thousands of years have passed away without any diminution of its magnitude, are we not irresistibly led to deny that there is in the moon any power of projecting a part of itself beyond the sphere of its own attraction?" "Extract from a Paper on the Meteoric Stones," *Trans. Amer. Philos. Soc.*, 1809, *6*: part 2, 401.

[24] James Ferguson, *Astronomy Explained upon Sir Isaac Newton's Principles . . .* , 3rd ed. (London, 1764), 5. On the vogue of Ferguson's work in American colleges, see Hornberger, *op. cit.*, 60.

break out all of a sudden with such lustre, it is by no means improbable that they are Suns whose Fuel is almost spent, and again supplied by some of their Comets falling upon them, and occasioning an uncommon blaze and splendor for some time: which indeed appears to be the greatest use of the cometary part of any system.[25]

Concerning the Milky Way and other nebulous appearances Ferguson had relatively little to say. The telescope, he declared, had failed to resolve the Milky Way into separate stars; its whiteness, as well as that of the so-called cloudy stars, must therefore be regarded as something more than blended starlight.

If the conception of the heavens presented in Ferguson's *Astronomy* persisted despite new discoveries and new ideas, it was not because the leading American astronomers were unaware of them. Rittenhouse's review of the progress and condition of astronomy before the American Philosophical Society in 1775 displayed a keen appreciation and substantial knowledge of developments on the other side of the Atlantic. He noted the growing opinion among astronomers that the orbits of the primary planets must gradually change their positions and offered his own conjecture that their poles would be found to revolve about a common center. He described Mayer's work on the apparent acceleration of the moon's motion and discussed the problems raised by it, observing that if the acceleration were real, "the destruction of this beautiful and stupendous fabric, may from thence be predicted with more certainty than from any other appearance in Nature." He then alluded to the researches of Alexander Wilson, professor of astronomy at the University of Glasgow, on sunspots and declared himself satisfied that they were permanent cavities in the body of the sun, some of them half as large as the earth. Turning his attention to the stars, he suggested that the Milky Way would provide a key to celestial mysteries:

. . . Millions of small stars compose it, and many more bright ones lie in and near it, than in other parts of heaven. Is not this a strong indication that this astonishing system of worlds beyond worlds innumerable, is not alike extended every way, but confined between two parallel planes, of *immeasurable,* though not *infinite* extent? Or rather, is not the Milky Way a vein of a closer texture, running through this part of the material creation? [26]

Rittenhouse was clearly moving with the times, but the textbooks of the day lagged somewhat behind. The second American edition of Ferguson's work, published in 1809 under the editorship of Professor Robert Patterson, had little to say about Herschel except for his discovery of a new planet and his observations of Saturn and its satellites. Of his theory of the heavens the only

[25] *Ibid.*, 367.

[26] David Rittenhouse, *An Oration Delivered February 24, 1775, Before the American Philosophical Society* . . . (Philadelphia, 1775), 25. This oration is reprinted in the "Appendix" to William Barton, *Memoirs of the Life of David Rittenhouse* . . . (Philadelphia, 1813), 543–577. Volume 2 of the American Philosophical Society's *Transactions*, 217–225, contains a letter from Mayer to Rittenhouse acknowledging his election to the American Philosophical Society and describing his method of studying the motions of the "fixed" stars. Rittenhouse's conjecture that the Milky Way is "not alike extended every way, but confined between two parallel planes" suggests that he was familiar to some degree with the speculations of Thomas Wright, Immanuel Kant, or Johann Heinrich Lambert (*Cosmologische Briefe über die Einrichtung des Weltbaues*, Augsburg, 1761).

suggestion was a passing remark in the glossary to the effect that some stars had been found to have a real motion and that this discovery had occasioned the conjecture "that not only the bodies belonging to the innumerable systems of stars are in motion round their respective centres, but that all the systems of bodies in the universe are themselves in motion round some common centre — and that thus they are prevented from approaching each other, which, from their mutual attractions, they must otherwise do." There was no mention of the nebular hypothesis. Instead, LaGrange's researches on the solar system were cited in qualification of Ferguson's argument against the eternity of the world. "M. de la Grange has demonstrated," Patterson declared editorially, "that the solar system is not *necessarily* perishable; but that the seeming irregularities in the planetary motions oscillate, as it were, within narrow limits; and that the world, according to the present constitution of nature, *may* be permanent." [27]

The first American edition of the Reverend Samuel Vince's *Elements of Astronomy*, published at Philadelphia in 1811, gave a fuller account of Herschel's work, including his idea of the generation of stars and the gradual formation and dissolution of star clusters. The disappearance of a star, it was conceded, might indicate the destruction of its system; likewise, the appearance of a new star might signify the creation of a new system of planets. The notion of "fixed stars" died hard, however. Both Vince's work and Patterson's edition of Ferguson used the expression, justifying it with the explanation that only a few of the stars had been shown to have a proper motion.

The successive editions of Jedidiah Morse's *American Universal Geography* provided another avenue by which news of Herschel's work reached the American reader. The edition of 1793 described his discovery of a seventh planet, but said nothing of his researches on nebulae or his speculations concerning the motion of the sun and stars. The "Introduction" to the edition of 1805, supervised by Professor Webber of Harvard, emphasized the vast increase in the number of known stars resulting from Herschel's improvement of the telescope. The Milky Way was described as consisting of innumerable separate stars, but the existence of nebulous stars was barely noted. The seventh edition, published in 1819, reported that Herschel had catalogued a great many nebulae and multiple stars and had located the solar system in the particular nebula called the Milky Way. The statement of previous editions concerning the immovability of the "fixed" stars was continued, apparently through carelessness, for a subsequent paragraph declared it the general belief that the fixed stars had their proper motions and noted Herschel's opinion that the solar system was moving toward a given point in the heavens. There was no mention, however, of Herschel's concept of stellar evolution or of the time scheme implied in it. Morse's world view was still essentially that of Ferguson: "If each of the fixed stars is a sun to a system, and every planet in all these

[27] James Ferguson, *Astronomy Explained upon Sir Isaac Newton's Principles . . .* , Robert Patterson, ed., 2nd Amer. ed. (Philadelphia, 1809), 116, n. Patterson also published *The Newtonian System of Philosophy; Explained by Familiar Objects, in an Entertaining Manner, for the Use of Young Ladies and Gentlemen,* the second edition of which appeared at Philadelphia in 1808.

systems is inhabited, like ours, with intelligent beings, as is supposable, what sublime ideas, what amazement, does such a view of the works of the infinite CREATOR inspire!"[28]

Most Americans shared Morse's faith in the stability of the heavens, but certain implications of the theory of gravitation and the corpuscular theory of light raised doubts in some minds. As early as 1751, four years before the publication of Kant's *Natural History and Theory of the Heavens*, Cadwallader Colden envisaged the creation and destruction of celestial systems in his *Principles of Action in Matter*. All particular systems in the general system of nature must eventually perish, reasoned Colden. In the case of the solar system, dissolution will result from the gradual diminution of the sun's energy by radiation. Falling comets may rekindle the solar furnace temporarily, but they cannot prevent its ultimate extinction.

> Nature, or more properly speaking, the infinite intelligent *Archeus*, has ordered so, that, since the several individual systems must in time fail, from their natural constitution, this defect is supplied by the generation of new and similar systems, the constant method of doing which is by fermentation, under the direction of the intelligent agent. So, supposing that all the comets, planets, and other parts of the solar system, by the failure of light in the sun, should at any time be united with the sun; then a chaos or confused mixture of the heterogeneous parts of matter must ensue; and thereby an extraordinary fermentation, during which the *Archeus* forms a new solar system, and a new heaven and a new earth may be produced. This conjecture seems to be confirmed by the appearing of new stars, and reappearing of some which before had disappeared.
>
> The duration of all the solar systems probably is infinite, in respect to the duration of any small system on this earth, whose period we know; and yet the duration of the solar systems may be infinitely small, in respect to the duration of the universe.[29]

Benjamin Franklin was inclined to suspect any theory of light which implied a steady attrition of the sun's mass. Newton's light corpuscles, he observed, gave no evidence of being able to move the slightest speck of dust, and the sun, for all its supposed loss of matter, seemed to preserve its ancient dimensions, and the planets their accustomed orbits. These objections stimulated James Bowdoin to attempt their removal in two communications to the American Academy of Arts and Sciences, of which he was the first president. Bowdoin began by showing why light particles, despite their velocity, could not displace dust motes. He then took up the objection that the corpuscular theory of light implied an eventual derangement of the solar system. Acknowledging that that system might eventually decay and require renovation, he declared it more likely that the Author of Nature had provided for its preservation, at least

[28] Morse, *American Universal Geography*, 7th ed. (Charlestown, Mass., 1819), *1*: 38. Herschel's work was known, of course, to all readers of the *Philosophical Transactions* of the Royal Society. Rittenhouse is known to have read his papers with great interest. The American Philosophical Society elected him a member shortly after his discovery of a new planet. In New England the Reverend Manasseh Cutler wrote Jeremy Belknap news of Herschel's discoveries as reported in the *Philosophical Transactions* for 1785, "lately come to hand." William P. Cutler and Julia P. Cutler, eds., *Life, Journals and Correspondence of Reverend Manasseh Cutler, LL.D.*, (Cincinnati, 1888), *2*: 238.

[29] Cadwallader Colden, *The Principles of Action in Matter* . . . (London, 1751), 167 (microfilm).

until such time as He saw fit to alter or abolish it. "Is it not conceivable," he asked, "that round the solar system, and the several systems, which compose the visible heavens, there might have been formed a hollow sphere, or orb, made of matter *sui generis*, or of matter like that of the planets, and surrounding the whole: having its inner concave surface at a proper distance therefrom; beyond which surface light could not pass, and between which, and the particles of light, there should be a mutual repulsion? And might not the Sun, or source of light, of each system, have been so placed, in respect of each other, and the concave surface of the surrounding orb, that there should be, by direct and repeatedly indirect reflections, an interchange of rays between them, in such a manner, as that to each there should be restored the quantity it had emitted; and thereby the waste of its matter be prevented: and this at the same time it dispensed its light to its particular system?" [30] Such a sphere, if properly constructed, might also serve to prevent the stars from being drawn together by gravity, since its own attractional force, varying with the density of its parts, would counterbalance the mutual attraction of the stars. Lastly, it might provide a residence for thousands or millions of creatures on both its interior and its exterior surfaces.

In his second memoir, Bowdoin supported his hypothesis with evidence from nature and from Scripture. The blue arch of heaven was the simplest proof, he declared. Still another was the whitish appearance of the Milky Way and of various other regions of the sky, suggesting the reflection of starlight from the inner surface of a sphere surrounding the sidereal systems. The evidences from Scripture, he acknowledged, were to be admitted only in a supplementary way. The Bible was not primarily intended to instruct man in physical science, but it nevertheless contains important hints concerning the general system of nature. It is not to be substituted for scientific investigation nor set in opposition to it, but wherever a probable scientific hypothesis is found to agree with the clear meaning of Scripture it becomes still more probable. "Such agreement, it is apprehended, shows the propriety and fitness of the interpretation: as, on the other hand, a disagreement with phenomena would prove the unfitness or falsity of any interpretation; and manifest it to be totally inadmissible." [31] Thus, for example, the distinction in various parts of the Bible between the heaven, the heaven of heavens, and the heavens of heavens seems plainly to confirm the idea of a series of concentric orbs each containing its own system of stars and planets. "When Scripture and phenomena thus agree, they mutually elucidate each other; and, in that case, what is deducible from the one, is confirmed by the other."

[30] James Bowdoin, "Observations on Light, and the Waste of Matter in the Sun and Fixt Stars, Occasioned by the Constant Efflux of Light from Them . . . ," *Mem. Amer. Acad. Arts Sci.*, 1785, *1*: 203. Franklin's argument was also answered by the Reverend Samuel Horsley in the *Philosophical Transactions* of the Royal Society, 1770, *60*: 417–440.

[31] James Bowdoin, "Observations Tending to Prove, by Phaenomena and Scripture, the Existence of an Orb, Which Surrounds the Whole Visible Material System . . . ," *Mem. Amer. Acad. Arts Sci.*, 1785, *1*: 230. Bowdoin notes a certain similarity between his conception and the discarded Ptolemaic theory of the heavens, but declares the resemblance quite accidental and superficial. He suggests that Whiston, "whose explanation of the *Mosaic* account of the creation is natural, and in general seems to be just," could have improved his theory by taking account of the Scriptural distinction between the upper and lower firmaments.

Bowdoin's argument suggests the extent to which the Christian doctrines of revelation and creation were involved in the discussion of astronomical matters. During the late eighteenth century the debate between the friends and foes of revealed religion reached a climax in the United States, coloring discussion of almost every topic. Deists like Ethan Allen and Tom Paine attempted to use astronomy to discredit Scripture and to establish Nature, conceived on the model of the solar system, as the one eternal, all-sufficient revelation of God. They ridiculed the crude astronomy of the Hebrews, compared the scientific prediction of eclipses and transits with the prophecies of Scripture, invoked the plurality of worlds against the doctrine of the fall and redemption of man, declared the impossibility of miracles in a law-bound system of matter in motion, and substituted the idea of the eternity of nature for the Biblical narrative of the creation. Accordingly, the friends of revelation, although they accepted the Newtonian universe in its grand outlines, were careful to guard against the anti-Christian inferences which deists sought to draw from it. The plurality of worlds was shown to be perfectly consistent with Scripture and with the attributes of the Creator. To Timothy Dwight, successor to Ezra Stiles at Yale, it was evident that "all worlds, and all intelligent beings, are parts of one kingdom of God," and hence that man, as an immortal creature, might justly hope to become acquainted with all parts of that kingdom in the course of eternity. Such knowledge would be unnecessary and impertinent in man's earthly condition but entirely appropriate in the Heaven of Heavens, "the place where all the works of God are studied, and understood, through an eternal progress of knowledge." [32]

On the question of the stability and wise contrivance of the heavens there was more agreement than disagreement between the friends and foes of revelation. Colden arrived at the idea of universal mutability in nature, but Tom Paine and Ethan Allen clung to the solar system as tenaciously as any creationist. "God, the great architect of nature, has so constructed its machinery, that it never needs to be altered or rectified," declared Allen, and Paine assured his readers that the motion of the planets is entirely different from the ordinary motions of terrestrial matter, since it operates "to perpetual preservation, and to prevent *any change* in the state of the system." Jefferson came eventually to acknowledge the evidences of mutability in the heavens, but, like his Christian adversary, Timothy Dwight, he managed to regard them as proofs of a superintending and renovating deity:

> Stars, well known, have disappeared, new ones have come into view; comets in their incalculable courses, may run foul of suns and planets, and require renova-

[32] Timothy Dwight, *Theology Explained and Defended in a Series of Sermons* . . . (New York, 1846), *1*: 286, 73. Rittenhouse, though somewhat deistically inclined, held views similar to Dwight's on the plurality of worlds. There was nothing in this conception, he declared, which precluded belief in the Christian doctrine of the atonement, since "infinite wisdom and power, prompted by infinite goodness, may throughout the vast extent of creation and duration, have frequently interposed in a manner quite incomprehensible to us, when it became necessary to the happiness of created beings of some other rank or degree." Like Dwight, too, he held out the hope that the Creator might be pleased to conduct man through "the several stages of his works" during the course of eternity (Rittenhouse, *Oration*, 26–27).

> tion under other laws; . . . were there no restoring power, all existences might extinguish successively, one by one, until all should be reduced to a shapeless chaos.[33]

In summary, it is evident that Americans took a lively interest in astronomy in the late eighteenth and early nineteenth centuries, keeping in touch with developments in Europe, publishing their own observations and theories, and reconciling as best they could the new discoveries and ideas with traditional conceptions of nature and man's place therein.

[33] Letter from Thomas Jefferson, Monticello, Va., to John Adams, 11 April 1823, quoted in A. A. Lipscomb, ed., *The Writings of Thomas Jefferson* (Washington, D. C., 1903–1904), *15*: 427. Compare Dwight, *op. cit.*, *1*: 134–135.

Science and the Public in the Age of Jefferson

*By John C. Greene**

THOMAS JEFFERSON'S interest in the sciences is well known, and it is usually assumed that his countrymen shared this interest to a considerable extent. According to Professors Krout and Fox, in their volume on *The Completion of Independence 1790–1830:*

> . . . Americans were keenly interested in natural science. Most of their knowledge, to be sure, came from Europe. In geology they debated the rival theories of James Hutton of Edinburgh and Gustav [sic] Werner of Freiburg; in biology the rival classifications of Linnaeus, Buffon and Cuvier. But it was socially significant that these ideas found a general hospitality. They were not imported for a small and artificial market, as they were into Russia by the exotic academy which Catherine the Great sponsored.[1]

How widespread was this "general hospitality" toward scientific ideas? How many Americans, and which Americans in particular, were "keenly interested" in the natural sciences? If there was general public interest in science, what were its sources? The present essay attempts to view these questions through the eyes of the American scientist of that period.[2]

That there was some spontaneous popular interest in science in these years cannot be denied. Dramatic events, such as Herschel's discovery of a new planet, Peale's exhuming of two mastodon skeletons, the appearance of a bright comet in 1807, and the fall of meteorites in Connecticut in the same year, were noticed in the public prints and occasioned considerable comment and speculation. There were also evidences of a more solid interest in science. Some of the general magazines of the period, such as *The Universal Asylum and Columbian Magazine* and *The Monthly Anthology and Boston Review*, were hospitable to science. The *Anthology*, for example, published Professor Benjamin Waterhouse's lectures on botany and noticed the progress of chemistry and mineralogy in Europe.[3] Charles Willson Peale's museum of natural history was maintained by the patronage of the Philadelphia public, and John Griscom's annual course of chemical lectures in New York was well subscribed from year to year.

* Iowa State College.

[1] J. A. Krout and D. R. Fox, *The Completion of Independence 1790–1830* (*History of American Life*, V) (New York, 1944), 313–314. See also William and Mabel Smallwood, *Natural History and the American Mind* (New York, 1941).

[2] The substance of this paper was presented at St. Louis on 28 April 1954, as part of a program on "Popular Science in American Culture," sponsored by the American Studies Association in connection with the annual convention of the Mississippi Valley Historical Association.

[3] Benjamin Waterhouse, "The Botanist," *The Monthly Anthology and Boston Review*, 1804, *1*: 390–393, 445–451, 492–496, 579–584; 1805, *2*: 9–14, 75–78, 124–128, 228–233, 347–352, 447–450, 503–508. On European science: 1809, *6*: 173–177, 280–282; 1811, *10*: 427–429. *The Universal Asylum and Columbian Magazine*, vols. 1–9 (Philadelphia, 1786–1792), has a considerable variety of scientific items, including extracts from standard European writers, reprints of American scientific essays, and a few original contributions, such as Captain Jonathan Heart's description and diagram of the aboriginal earthworks on the Muskingum. There is also considerable material of scientific interest in *The American Magazine. . .* , Noah Webster, ed., Nos. 1–12 (New York, 1787–1788). Dennie's *Port Folio*, the most successful general magazine of the period, paid little attention to science. Many magazines, including the *Monthly Anthology*, borrowed items of scientific intelligence from the *Medical Repository* after its founding in 1798.

These examples of popular interest in natural science should not be overlooked, but they must not be taken as typical, nor as proving that science had a broad popular appeal. Quite the contrary was the case, if we can believe the American scientists of the period, struggling, as they were, to create a place for science in the thoughts of their countrymen. "I was compelled to clear the ground to prepare it before I sowed the seed," said Benjamin Waterhouse. "I found I must first excite a taste, and then try to gratify it." [4] The same was true of other sciences. "Chemistry," wrote Benjamin Silliman in 1810, "is here almost a new pursuit, and hence it is not uncommon to find even intelligent men manifesting an entire ignorance of its nature and utility." [5] Popular interest in science was something to be wished for and worked for, not something to be taken for granted.

It should not be thought that America was greatly different from Europe in this respect, however. The period under discussion was a brilliant one in the history of Western science — the age of Laplace and Herschel, Lavoisier and Davy, Hutton and Werner, Cuvier and Lamarck — but it was not a great age of *popular* science. The earlier enthusiasm for popularizing Newtonian physics and astronomy was on the wane, and the day of the mechanics' institute, the lecture circuit, and the popular scientific press had not yet arrived on either side of the Atlantic. In France, one could point to the public lectures at the Museum of Natural History; in Britain, to Davy's lectures at the Royal Institution, to Accum's chemical school in London, to the attendance of workingmen at lectures on experimental physics at the University of Glasgow, and to the gradually increasing number of introductory works on natural philosophy, some of them frankly popular in character.[6] But these were only the beginnings of a movement which was not to come into its own until the second quarter of the nineteenth century, by which time the progress of urbanization and industrialization, the spread of public education, and the growth of a strong will to self-improvement among the artisan class had created the conditions requisite for large-scale popularization of science.

The flowering of popular science requires more than a potential audience, however. It presupposes also a certain degree of maturity in science itself and some sense of security and leisure on the part of the scientist. It was in this respect that conditions in the United States were distinctly less favorable to the growth of popular science than those in Europe. "Our elder brethren in Europe know not the difficulties that the first settlers in science have to encounter," wrote Benjamin Waterhouse in 1811.[7] What were some of these difficulties? In the first place, very few Americans were in a position to devote themselves whole-heartedly to scientific research. Science was an avocation

[4] Letter from Benjamin Waterhouse to J. E. Smith, Cambridge, Mass., 24 July 1811, quoted in Lady Smith, ed., *Memoir and Correspondence of the Late Sir James Edward Smith, M.D.* (London, 1832), II, 174.

[5] Benjamin Silliman, "Notes to the . . . American Edition of Henry's Chemistry," in William Henry, *An Epitome of Experimental Chemistry* . . . 2nd American ed. from the 5th English ed. (Boston, 1810), i. Silliman continues: "Happily the increasing taste for the science, which is indicated by our augmenting sources of chemical information, gives good grounds to hope that this species of knowledge will soon be extensively diffused."

[6] E. T. Hamy, "Les Derniers Jours du Jardin du Roi et la Fondation du Muséum d'Histoire Naturelle," *Centenaire de la Fondation du Muséum d'Histoire Naturelle* (Paris, 1893), 3–162; Daniel Mornet, *Les Sciences de la Nature en France, au XVIIIe Siècle* . . . (Paris, 1911). A. Laming, "The Origins of the Popularization of Science," *Impact of Science on Society*, 1952, *3*: 233–258. A. E. Dobbs, *Education and Social Movements 1700–1850* (London and New York, 1919), 141–142, 170–172; D. M. Turner, *History of Science Teaching in England* (London, 1927), 52–62. C. A. Browne, "The Life and Chemical Services of Frederick Accum," *Journal of Chemical Education* (1925), 1–58.

[7] Letter from Benjamin Waterhouse to J. E. Smith, 24 July 1811, quoted in Smith, ed., *Memoir and Correspondence*, II, 174.

for most of those who loved it, and even those who taught natural philosophy or natural history in the colleges were hard pressed with other duties. Barton, Mitchill, and Waterhouse were primarily practitioners and teachers of medicine, and only secondarily teachers and researchers in natural history. Mitchill was also active in politics. The astronomers David Rittenhouse and Andrew Ellicott were kept busy with surveying, clock-making, and other activities to support their families. Muhlenberg, Madison, and Cutler were clergymen. It was much to be regretted, declared Benjamin Smith Barton, "that the principal cultivators of natural science, in the United States, are professional characters, who cannot, without essentially injuring their best interests, devote to these subjects that sedulous attention which they demand." [8]

When time for scientific research was available, the necessary tools were frequently lacking. "The study of natural history in this country," wrote Manasseh Cutler to a European correspondent in 1799, "is in its infancy. . . . This deficiency has been, in part, owing to the great scarcity of books on natural history. . . . We have many public libraries, consisting of large and well chosen collections of books from Europe, excepting on the different branches of natural history, in which there are few, and those mostly ancient authors who wrote before the Linnaean system was formed, and our booksellers import no books on this subject. . . . We have no cabinets of natural history in America, excepting one in Philadelphia and another in Boston. These consist of small collections without any systematic arrangement. They are kept merely for the purpose of getting money by showing them to common people, and consist principally of exotics." [9] In like vein, Benjamin Silliman, a decade later, lamented the scarcity of chemical apparatus in the United States and the inability of American industry to supply the want. "Under such circumstances," he observed, "it is easy to see, that the practice of chemistry, even for philosophical purposes, must be attended with difficulties of no small magnitude. . . ." [10] Under such circumstances, it may be added, the energies of American scientists could not be devoted to popularization except in so far as the creation of an intelligent appreciation of science was necessary to its very existence.

The place of efforts at popularization in the larger struggle to put American science on its feet may be illustrated in various ways. The *Catalogue of All the Books Printed in the United States*, issued by the Boston booksellers in January, 1804, provides a convenient starting point. Of the 1,338 titles listed in this catalogue, not more than twenty can be considered works of science, if medical titles are excepted.[11] Judging from the titles listed and from the

[8] Benjamin Smith Barton, "Miscellaneous Facts and Observations," *Philadelphia Medical and Physical Journal*, 1805, *1*: 158–159. See also the "Introduction" to Barton's *Fragments of the Natural History of Pennsylvania*, Part I (Philadelphia, 1799).

[9] Letter from Rev. Manasseh Cutler to Gustav Paykull (Hamilton, 14 February 1799) quoted in W. P. and Julia Cutler, eds., *Life, Journals and Correspondence of Reverend Manasseh Cutler, LL.D.* (Cincinnati, 1888), II, 298. As late as 1821, Thomas Jefferson, protesting the fifteen per cent tariff on books imported into the United States, asserted that "the home demand is not sufficient to justify the reprinting any but the most popular English works, and cheap editions of a few of the classics for schools. Of many important books of reference there is not perhaps a single copy in the United States; of others but a few, and these too distant often to be accessible to the scholar generally." Jefferson to ———, Monticello, 28 September 1821, quoted in A. A. Lipscomb, ed., *The Writings of Thomas Jefferson* (Washington, D. C., 1903–1904), XV, 338–339.

[10] Benjamin Silliman, "Notes to . . . Henry's Chemistry," in Henry, *op. cit.*, ii.

[11] *Catalogue of All the Books Printed in the United States, With the Prices, and Places Where Published, Annexed* (Boston, 1804), reprinted in Adolph Growoll, *Book Trade Bibliography in the United States in the Nineteenth Century* (New York, 1939), 1–79. Under the heading "Physic" are listed, besides medical books, works on chemistry by Parkinson, Lavoisier, Penington, and Henry; works on natural philosophy by Enfield, Rumford, and Nicholson; two works on astronomy and related subjects; Erasmus

notices in the *Medical Repository* in subsequent years, chemistry was the most popular science of the day. "Elementary books on chemistry are become almost as common as books on spelling and arithmetic," the editors of the *Repository* observed in 1808. "The compilations of Parkinson, Thompson, Henry, Accum, Jacobs, Murray, Lagrange, Spalding, Ewell, and several others which have been published within a few years, evince the prevailing taste for this interesting science." [12] As the list of authors suggests, most of these books were reprints of foreign works. Few of them were truly popular in character. The two which ran to several American editions before 1820 — James Parkinson's *Chemical Pocket-Book* and William Henry's *Epitome of Experimental Chemistry* — were technical manuals for the beginning student. Parkinson described his book as "an agreeable pocket companion for the lovers of Chemistry in general; and more particularly so for those who may be just engaging in the study of this most useful and interesting science." [13] Works of a more popular character, such as Madame Marcet's *Conversations on Chemistry* or Dr. Thomas Ewell's *Plain Discourses on the Laws or Properties of Matter*, were few and far between. "The common herd of philosophers seem to write only for one another," complained Thomas Jefferson in a letter acknowledging receipt of Ewell's work. Justice Bushrod Washington was equally plaintive in the same connection: "I have not met with a single treatise which has not appeared unnecessarily obscured by technical terms, which only scholars can understand. They have been more generally addressed to the comprehension of professional and learned men, than to those of the humble walks of life, for whose use this science might be made most essentially to contribute, by adapting it to their capacities, and by pointing out the way by which its principles may be applied to the more common arts, in which they are daily employed." [14] The same might have been said with equal truth of the current output of books dealing with the other branches of natural philosophy.[15]

Darwin's *Botanic Garden*, *Temple of Nature* and *Zoonomia*; Goldsmith's *History of the Earth and Animated Nature*; Lavater's *Physiognomy*; and several books on natural theology by St. Pierre, Paley, and Sturm. Concerning the book-selling trade in New York, Dr. Samuel Latham Mitchill wrote in his *Picture of New York* (New York, 1807), 139–140: "The high price of paper, labour and taxes in the British islands, has been very favourable to authorship, and the publication of books at home. Foreign publications, too, come to us charged with a duty, in our own ports, of 15 per cent in addition to other expenses. To encourage the domestic manufacture of paper, Congress, in 1804, exempted from impost, all imported foreign rags. This has given a lively spring to the whole complicated manufacture, of paper from rags, and of books, pamphlets and gazettes from paper. Hence it happens that authors find it easy to publish their original works; and editors proceed with equal facility to re-print foreign books. . . . With the increase of population, there is at least a proportional increase of reading. The ratio may be calculated even greater from the democratic temper of the people."

[12] Review of *Conversations on Chemistry* (Philadelphia, 1806), in *The Medical Repository*, 1808, *11*: 50.

[13] James Parkinson, *The Chemical Pocket-Book; or Memoranda Chemica: Arranged in a Compendium of Chemistry: With Tables of Attractions, &c. Calculated as Well for the Occasional Reference of the Professional Student, as to Supply Others with a General Knowledge of Chemistry.* 1st American ed. from the 2nd London ed. (Philadelphia, 1802), vii. For a description of early chemistry books, both American and foreign, see Edgar Fahs Smith, *Old Chemistries* (New York and London, 1927); also his *Chemistry in America* (New York and London, 1914).

[14] Letters from Thomas Jefferson, Monticello, August, 1805, and from Bushrod Washington, Mt. Vernon, 15 September 1805, to Thomas Ewell, quoted in Thomas Ewell, *Plain Discourses on the Laws or Properties of Matter Containing the Elements or Principles of Modern Chemistry; With More Particular Details of Those Practical Parts of the Science Most Interesting to Mankind, and Connected with Domestic Affairs* (New York, 1806), "Preface," 8–9.

[15] An advertisement dated May, 1809, in the New Haven edition of Madame Marcet's *Conversations on Chemistry*, lists ten works on chemistry and as many on astronomy and physics, including Ferguson's *Astronomy*, Euler's *Letters*, Rumford's *Essays*, Cavallo's *Electricity*, and Noah Webster's *Elements of Natural Philosophy, With Notes and Corrections by Professor Robert Patterson of the University of Pennsylvania.* Patterson also edited two American editions of Ferguson's *Astronomy*, and two of a popular work entitled *The Newtonian System of Philosophy; Explained by Familiar Objects, in an Entertaining Manner*, in which Master Tom Telescope diverts and instructs a

Natural history was in a similar position, except that books in this field were less numerous. Botany was the favorite science here. In the popular vein, Benjamin Waterhouse's botanical lectures, the same which had appeared in the *Monthly Anthology*, were published in book form in 1811. It had been his original idea, Waterhouse explained in the "Advertisement," to include lectures on other branches of natural history too, but he had relinquished the idea on the plea of the editors "that mineralogy would be less popular than botany; and therefore less adapted to such a monthly magazine of knowledge and pleasure, as the Anthology was meant to be; and less likely to attract the attention and patronage of readers of both sexes." [16] Waterhouse had also been influenced by the knowledge that the merchants of Boston were being asked to contribute to the establishment of a botanical garden at Harvard, yet few of them knew enough of botany to understand the purposes which such a garden might serve. These facts were set forth, he said, in order that the European disciples of Linnaeus might "see the reason, and therefore excuse the popular dress, in which BOTANY, that beautiful handmaid of Medicine, has been introduced to the inhabitants of a region characteristically called by the English a century ago, THE WILDERNESS." [17]

The other botanical works which issued from the American press in this period could scarcely be described as popular. Benjamin Smith Barton's *Elements of Botany* was a textbook designed for the use of his students at the University of Pennsylvania. Dr. Jacob Bigelow's *Florula Bostoniensis* was produced in response to popular acclaim of his botanical lectures in Boston, but it was little more than a catalogue of plants in the vicinity of Boston. For his Harvard students, Bigelow undertook to edit an American edition of Sir James Edward Smith's *Introduction to Physiological and Systematic Botany.* "Our Botanists," he wrote Smith, "are not yet sufficiently numerous to induce the booksellers to publish large works; but as the country grows I hope the taste for science will increase." [18] The works of Linnaeus were in demand among American botanists, but none appears to have been reprinted in the United States. Nor do the more popular works of Buffon, although available in several English translations, seem to have attracted American booksellers. The favorite European writer on natural history, judging from the number of American editions of his writings, was "the celebrated Dr. Darwin," as Erasmus Darwin, grandfather of the immortal Charles, was known. His poetical renditions of Linnaean botany were required reading for ladies of fashion. *The Botanic Garden* and *The Temple of Nature*, both of which appeared in American dress,[19] acquainted the young lady who read them with the "sexual system" of botany, but they did much more. They introduced her to some of the latest discoveries and most advanced speculations in the world of science: the geological uniformitarianism of Hutton, the brilliant discoveries and conjectures of Herschel, the notion of the mutability of species, and the

---

group of young ladies and gentlemen gathered at the house of Mr. Setstar with lectures on the solar system, the atmosphere, animals, and man.

[16] Benjamin Waterhouse, *The Botanist. Being the Botanical Part of a Course of Lectures on Natural History, Delivered in the University at Cambridge* . . . (Boston, 1811), "Advertisement," vi.

[17] *Ibid.*, viii.

[18] Letter from Jacob Bigelow to J. E. Smith, Boston, 18 November 1813, quoted in Smith, ed., *Memoir and Correspondence*, II, 187. See also the "Preface" to Bigelow's *Florula Bostoniensis* (Boston, 1814), page v: "The common standard works of the science . . . are hardly so much as heard by name in our bookstores. These works, even when obtained, being principally in Latin, are useless to a great class of amateurs of the science, who are not conversant in the learned languages."

[19] The first American edition of *The Botanic Garden* was published in New York in 1798; an abridged version, *Beauties of the Botanic Garden*, followed in 1805. *The Temple of Nature* was published in both Baltimore and New York in 1804.

conception of human history as a slow progress from savagery. A few pages into the first canto of *The Botanic Garden*, for example, she would encounter Darwin's explosive version of the creation:

"LET THERE BE LIGHT!" proclaim'd the ALMIGHTY LORD,
Astonish'd Chaos heard the potent word;
Through all his realms the kindling Ether runs,
And the mass starts into a million suns;
Earths round each sun with quick explosions burst,
And second planets issue from the first . . .[20]

A glance at the footnote would apprise her that Mr. Herschel supposed the stars to be moving around some common center, and prepare her for Darwin's argument that the projectile force which prevented the stars from rushing together must have originated in a gigantic explosion in which "the whole of Chaos, like grains of gunpowder, was exploded at the same time, and dispersed through infinite space at once, or in quick succession in every possible direction," while subsidiary explosions in the various stars projected planetary and cometary systems into motion around them. If she cared to consult the "Additional Notes" at the end of the volume, she would find a fuller exposition of the author's theory of the genesis of solar systems from "sunquakes" and a comparison of it with Buffon's hypothesis, much to the disadvantage of the latter.

Darwin was a popularizer of high calibre, but the fact remains that the work from his pen which ran through more American editions than any other was not popular in character. It was his *Zoonomia: Or the Laws of Organic Life*, the book which contains his famous hypothesis of organic evolution.[21] This fact would seem to bear out the thesis that the market for scientific books in the United States in this period was largely among serious students of the sciences. Probably the potential audience was fairly well described by the anonymous authors of *The Epitome of Electricity and Galvanism*, published at Philadelphia in 1809, when they wrote: "Our views will be fully answered, if [our work] shall be found well adapted to assist youth in their academical and philosophical studies, and at the same time, to afford amusement to men of learning, and some useful information to gentlemen of leisure." [22]

This description would apply equally well to the public which read scientific periodicals. The various scientific societies, notably the American Philosophical Society and the American Academy of Arts and Sciences, published their transactions and memoirs from time to time, but these publications had a very limited circulation.[23] The short-lived *Emporium of Arts and Sciences*, pub-

[20] Erasmus Darwin, "The Botanic Garden. A Poem with Philosophical Notes," in *The Poetical Works of Erasmus Darwin* . . . (London, 1806), I, 9–11, Canto I, l. 103 ff.

[21] Part I of Darwin's *Zoonomia* was published in New York in 1796; Part II, in two volumes, appeared in Philadelphia in 1797. Other American editions followed in 1803 (Boston), 1809 (Boston), and 1818 (Philadelphia). The high esteem in which Darwin was held by American medical men is revealed in Dr. Charles Caldwell's "Introductory Address" to the Philadelphia edition of 1797, in which Darwin's *Zoonomia* is praised as "a splendid and towering *monument* of genius, destined long to survive its illustrious builder," and *The Botanic Garden* is described as "sufficient alone to immortalize his country and his age!" Samuel Latham Mitchill and Benjamin Waterhouse were also ardent admirers of Darwin. Joel Barlow's *Columbiad* shows the influence of Darwin on American literature.

[22] [Jacob Green and Erskine Hazard], *An Epitome of Electricity and Galvanism. By Two Gentlemen of Philadelphia* (Philadelphia, 1809), "Preface." The identity of the authors was subsequently revealed. Both had recently graduated from the University of Pennsylvania when their book appeared. In 1818 Jacob Green was appointed professor of chemistry at the College of New Jersey, of which his father was president.

[23] Concerning the second volume of the *Memoirs* of the American Academy of Arts and Sciences, published in 1804, the editors of the *Medical Repository* wrote in 1808: "Although this publication of one of the most respectable institutions in our nation, came forth about

lished by John Redman Coxe and Thomas Cooper, was devoted to the technological applications of science, but was not a popular science magazine in the modern sense. "I will not," Cooper announced on taking over the editorship in 1813, "condescend to make this a work of mere amusement, for the purpose of sale — one that shall suffice merely, under the show of science, to enable the reader to trifle away an hour, and to skim the surface of a great many subjects for the purpose of a superficial and conversation knowledge. Many pages of this work to a general reader will be very dull; but it will be my fault if they are not useful to those who read for improvement." [24] Apparently the number of such readers was not large, for the magazine was soon discontinued.

The only thing approaching a general scientific magazine was the *Medical Repository*, published in New York by Dr. Samuel Latham Mitchill and associates. The *Repository* was by no means simply a medical journal in its early years. The preface to the first volume, issued in 1798, declared the journal to be a "depositary of facts and reasoning relative to Natural History, Agriculture, and Medicine." In subsequent prefaces the editors announced their intention "to notice every leading fact, and every important improvement in the progress of the physical sciences both in Europe and America." They were pleased, they said, to find the scope and circulation of the publication increasing, but they had made no money in the enterprise, "and every prospect of that kind is at present so remote and uncertain, as to offer no encouragement of such expectations." Instead, their reward must be "the consoling and animating reflection that the importance of the objects we pursue is not limited to the place in which we reside, nor to the times in which we live." [25]

The *Repository* posted its readers on scientific developments on both sides of the Atlantic. From its pages they learned not only of Cuvier's work on fossil quadrupeds and Davy's researches in chemistry but also of Maclure's geological explorations, of Peale's exhuming of the mastodon, of the appearance of the successive volumes of Wilson's *American Ornithology*, of the publication of new scientific journals in the United States, and of many other undertakings and accomplishments. All these were reported with great enthusiasm for science, and especially for American science. "Too long," said the editors, "has it been fashionable for our people to seek scientific news from transatlantic regions, while they neglected the manifold novelties by which they were surrounded at home. But latterly a more correct opinion has prevailed among them . . . to turn their backs to the east, and direct their views to the inviting and productive regions of the interior of America." [26] The day was at hand, the editors declared, when the business man, the gentleman, and the finished scholar must all pay greater attention to natural history: "How boorish . . . is it to be ignorant of the general history of the elephant!" Thus did Mitchill and his colleagues seek to promote the cause of science, and particularly of American science, but their approach was scholarly, their audience limited, and their reward uncertain.

The demand for public lectures on science was correspondingly slight. The

---

three years ago, yet so slowly do works of this kind travel from city to city, that we never until lately procured a sight of it. The difficulty of intercourse between one part of our nation and another, surpasses, as we have had repeated cause to remark before, any thing we experience in respect to the British islands, and, we might add, some other parts of Europe." *Medical Repository*, 1808, *11*: 273.

[24] *The Emporium of Arts and Sciences*, n.s., I (1813), "Prospectus," 2. *The Emporium* was conducted by Dr. John Redman Coxe in 1812; it continued under Cooper's editorship during 1813 and 1814, after which it expired.

[25] Prefaces to the *Medical Repository*: 1798, *1*: iv; 1803, *6*: viii; 1805, *8*: vii; 1806, *9*: iii–iv.

[26] *Medical Repository*, 1808, *9*: 42; 1804, *7*: 403.

only person who earned a substantial part of his living by popular scientific lectures during this period was John Griscom.[27] About 1795, while teaching school in New Jersey, Griscom became interested in chemistry through reading a translation of Lavoisier's *Elements of Chemistry*. He went to Philadelphia to hear Professor Woodhouse lecture, ordered apparatus from London, and soon was teaching chemistry in the school at Burlington. Upon removing to New York in 1800, he tried his hand at lectures by public subscription. They were an immediate success, so much so that Griscom rented part of the Friends' graveyard on Liberty Street and built there a substantial brick building to house his operations. He continued there until the City authorities donated the Old Alms House for use by Griscom and by John Scudder, keeper of Scudder's Museum.[28] Griscom's lectures thus became a familiar institution in New York, attended by humble mechanics as well as ladies of fashion, noticed from time to time in the newspapers, and mentioned with respect in Mitchill's *Medical Repository*. The lectures of 1807–1808 were attended, said the *Repository*, by "upwards of one hundred persons of both sexes, and have obtained a distinguished degree of approbation." Several of Griscom's auditors that winter volunteered to inhale nitrous oxide, or "laughing gas," to demonstrate its effects. One person fainted, but persons in good health, especially those inclined to corpulency, exhibited "the highest degree of exhilaration and rapture." [29]

Griscom did not confine himself to chemistry. In 1811 he announced that his chemical lectures would be preceded by a full course on natural philosophy, including astronomy. The topics discussed and the fashionable character of his audience are suggested in Fitz-Green Halleck's poem "Fanny," in which the heroine is pictured as one of the "bright-eyed maids and matrons" who attended "Griscom's conversationes," and there learned:

> Words to the witches in Macbeth unknown,—
> Hydraulics, Hydrostatics, and Pneumatics,
> Dioptrics, Optics, Katoptrics, Carbon,
> Chlorine, and Iodine, and Aerostatics;
> Also,—why frogs, for want of air, expire;
> And how to set the Tappan sea on fire.

Griscom's career continued well into the age of popularization in the second quarter of the nineteenth century. Along with Silliman he could claim the honor, as Silliman put it, "of inaugurating a system which has since become almost universal in the United States."

Griscom was the exception rather than the rule in the opening decades of the century, however. Most public lectures on science before 1815 were one-season performances before small audiences, often without recompense to the speaker. In the winter of 1807–1808 Boston society turned out to hear a visiting Frenchman, one Monsieur Godon, lecture on mineralogy. In the spring of 1809 chemistry was the topic, the lecturer Dr. John Gorham, recently returned from his studies in Europe. "Although the subject was a novel one, and though

[27] This account of John Griscom is based primarily on Edgar F. Smith's *John Griscom 1774–1852 Chemist* (Philadelphia, 1925).

[28] The *Medical Repository* (3rd Hexade, 1811, *2*: 88) refers to John Scudder as "a man who, by his own native impulse, has conceived an ardent desire of collecting, and who possesses uncommon skill in preserving, the material of natural history." Among the exhibits in his museum, the *Repository* drew special attention to his white bear of Greenland, "his royal tygress of Hindostan, and his collection of American squirrels"; also to his pelicans, gannets, loons, and owls, his collection of insects, zoophytes, and serpents, and his small but growing collection of shells.

[29] *Medical Repository*, 1808, *11*: 320.

few persons in this metropolis had cultivated a taste for it," reported the *Monthly Anthology*, "yet his lectures were as fully and constantly attended, as the scale upon which his modesty had induced him to commence them could admit." [30] The response to the botanical lectures of Waterhouse and Bigelow was equally gratifying.

So it was in other cities. Albany reported a course of chemical lectures attended by about sixty persons, including many ladies. Philadelphia was much taken with the botanical lectures of Joseph Correa de Serra, Portuguese minister to the United States and a botanist of international reputation. In Wilmington, Delaware, William Baldwin found great enthusiasm for botanical instruction among the ladies of the city, and Dr. John Vaughan complimented them for their faithful attendance at his lectures on natural philosophy. In Charleston, South Carolina, Stephen Elliott lectured on natural history to the Literary and Philosophical Society.[31] These various public lectures evinced a growing interest in the sciences, an interest compounded of native curiosity and civic pride, but they were small affairs compared to Silliman's grand performances at the Lowell Institute a quarter of a century later.

A final illustration of the effort to create and exploit popular interest in science is to be found in the story of Peale's Museum, established in Philadelphia in the years 1786–1809 by the well-known portrait painter Charles Willson Peale. Of all the museum entrepreneurs of the period, and there were a considerable number, Peale alone seems to have had a vision of the service which a popular museum of natural history could render both to the progress of science and to the scientific education of the general public. His ideal was that of a great museum publicly supported as part of a state or national university, containing within its walls a "world in miniature," its exhibits open alike to the student of science and to the populace, its program embracing not only the display of the productions of nature but also the instruction of the public by lectures and guided tours and the linking of science to the arts of music and painting. Such a museum would be a veritable "Temple of Wisdom," a place where men would learn to imitate the benevolence of the Creator by observing the order and beauty of his creation.[32]

If the reality fell short of the ideal, it was not for lack of effort on Peale's part. He labored constantly to extend and improve his collections and to arrange them scientifically. He studied the methods of European museums and exchanged specimens with them. He undertook to produce a catalogue of the exhibits in his museum and to lecture on natural history. He tried again and again to secure public support for the institution, and, when this was not

[30] Letter to the Editors, *Monthly Anthology*, 1809, *6*: 236–237. Gorham, a graduate of Harvard College and Harvard Medical School, was appointed Adjunct Professor of Chemistry and Materia Medica at Harvard in 1809. While abroad, he attended the popular lectures and demonstrations of Frederick Accum, manufacturer of chemical apparatus and a popular writer and lecturer on chemistry. See I. B. Cohen, *Some Early Tools of American Science* . . . (Cambridge, Massachusetts, 1950), 81 ff.

[31] *Medical Repository*, 1801, *4*: 291; 3rd Hexade, 1811, *2*: 89; William Darlington, ed., *Reliquiae Baldwinianae*: *Selections from the Correspondence of the Late William Baldwin, M.D.* (Philadelphia, 1843), 32, 171–172.

[32] The present account of Peale's activities and ideas draws heavily on Charles C. Sellers, *Charles Wilson Peale* (Philadelphia, 1947). There is a contemporary description of Peale's Museum in the *Port Folio*, 1807, *4*: 293–295; also in Dr. James Mease's *Picture of Philadelphia* (Philadelphia, 1811), 311–314. On American museums generally, see William M. and Mabel S. Smallwood, *Natural History and the American Mind* (New York, 1941), Chaps. 4–5. The *Medical Repository*, 3rd Hexade, 1811, *2*: 88–89, describes briefly Scudder's Museum in New York and Trowbridge's Museum in Albany. The *Monthly Anthology*, 1804, *1*: 143, carries a notice of the Columbian Museum in Boston, established in 1795 by Daniel Bowen, destroyed by fire in January, 1803, but reestablished eight months later. Besides paintings, wax figures, and the like, the museum boasted a carved replica of Peale's mammoth: "Which may be justly ranked among the *greatest* artificial Curiosities."

forthcoming, to make the museum pay its way without loss of scientific integrity. The exhuming of two mastodon skeletons in New York state was his supreme effort both at extending the bounds of scientific knowledge and at making his museum self-supporting. It served both purposes admirably; the mounted skeletons attracted national and even international attention in both the scientific and the non-scientific world.[33] Yet even in this hour of triumph Peale did not lose sight of his ultimate objective, the transformation of his enterprise into a public institution. In this he failed, however. The Jeffersonian party to which he gave allegiance was too much attached to the principles of economy in government and strict construction of the Constitution to approve federal expenditures for a national museum, and the Pennsylvania legislature, its Philadelphia members included, would do nothing more than donate the upper rooms of the old State House for the housing of the museum. The legislators, like most of the visitors who paid their twenty-five cents to see the exhibits, regarded the museum as a kind of show; they were impressed with Peale's financial success but took it as proof that his museum did not need public support.

After Peale's retirement in 1809, the precarious balance between scientific integrity and pecuniary profit was tilted dangerously in the practical direction. In the hands of Rubens Peale, the "Temple of Wisdom" became increasingly a business operation, hard pressed by competition from the numerous rivals to which its financial success had given rise. Peale's vision of the museum as a great instrument for the promotion and popularization of science was not to be realized in his own day nor for many decades to come. Jefferson probably appraised the situation correctly when, in 1807, in response to a request for his subscription to a museum venture in Williamsburg, he wrote:

> In the particular enterprises for museums, we have seen the populous and wealthy cities of Boston and New York unable to found or maintain such an institution. The feeble condition of that in each of these places sufficiently proves this. In Philadelphia alone, has this attempt succeeded to a good degree. It has been owing there to a measure of zeal and perseverance in an individual rarely equalled; to a population, crowded, wealthy, and more than usually addicted to the pursuit of knowledge.[34]

It appears, then, (1) that popular interest in science in this period, though by no means completely absent, was inadequate to provide solid support for American science, and (2) that such public interest as existed was due in no small part to the efforts of the devotees of science to convince their countrymen of the importance and utility of scientific research. It may be worth while, therefore, to glance at the types of appeal which were used in this campaign for public support.

The patriotic appeal was the most common and probably the most effective. Americans took pride in the size, resources, and varied natural attractions of their country and expected, of course, that these material advantages would be matched by accomplishment in the arts and sciences. Jefferson won the everlasting gratitude of his countrymen when he undertook to refute Buffon's suggestion that the species of animate nature in the New World were smaller and less vigorous than those of the Old, perhaps because the western hemisphere

[33] The excitement caused in European scientific circles by Peale's assembling of two nearly complete skeletons may be judged from Cuvier's remarks in his memoir, "Sur Le Grand Mastodonte," *Annales du Muséum d'Histoire Naturelle*, 1806, *8*: 270–312.

[34] Letter from Thomas Jefferson to G. C. de la Coste, Washington, 24 May 1807, quoted in Lipscomb, ed., *The Writings of Thomas Jefferson*, XI, 206–207.

had been raised from the waters at a later date. The bones of the mammoth and the megalonyx were hailed as proof positive that Buffon was wrong, and the gorge of the Niagara Falls was adduced in favor of the very respectable antiquity of nature in America. Travel accounts and state histories called attention to the natural environment and to the importance of studying it. Jeremy Belknap devoted the third volume of his *History of New Hampshire* to the natural history of the state, and Samuel Williams made the most of his scientific training in his *Natural and Civil History of Vermont.* John Drayton's *View of South Carolina in her Natural and Civil Concerns* described the giant teeth and bones dug out of Biggin Swamp in 1795 and the great banks of enormous oyster shells along the Santee River. Even municipal guidebooks were used to acquaint the public with natural history. Samuel Latham Mitchill's *Picture of New York* included several pages of geological description extracted from the "Mineralogical Sketch" which Mitchill had published earlier in the *Medical Repository.* Dr. James Mease introduced the readers of his *Picture of Philadelphia* to Lewis Evans' theory of the origin of the Atlantic coastal plain and to various speculations concerning changes in the climate of Philadelphia. Dr. Daniel Drake's *Picture of Cincinnati and the Miami Country* contained chapters on physical and medical topography and on the "Indian mounds" of the Mississippi Valley; the "Appendix" treated of earthquakes, the aurora borealis, and the southwest wind.

All in all, considerable attention was paid to natural history in these various kinds of books, either because the reading public was interested in such matters or because the authors of the books thought that their readers ought to be interested in them. Probably both reasons applied in varying degree. The editors of the *Medical Repository* reprimanded the Reverend Thaddeus Mason Harris for the paucity of scientific information in his *Journal of a Tour into the Territory Northwest of the Alleghany Mountains.* They suggested that he might have written a more interesting book if he had been "duly prepared by having studied mineralogy, geology, and other physical sciences."[35] Volney's *View of the Soil and Climate of the United States* was more to their liking, but the translator and editor of the American edition of this work, Charles Brockden Brown, drew the line at Volney's suggestion that the American government establish a society of linguists to collect and analyze Indian vocabularies. "The American citizen," said Brown, "will smile at this proposal. The *great* importance here bestowed on the business of collecting the dialects of barbarous tribes, who are hastening to oblivion, for the sole purpose of throwing a faint light on the question whether these tribes originally came from the north of Asia, will hardly be felt by the busy merchant, artizan, or farmer, or by their public representatives . . ."[36]

The patriotic and the utilitarian appeal for the support of science were closely intertwined. Science was recommended to the public as the example *par excellence* of useful knowledge: it would enrich the individual, make the nation great in peace and war, and advance mankind on the road to happiness. It could

[35] *Medical Repository*, 1806, *9*: 398. Harris' *Journal* was not devoid of scientific interest, however. The *Monthly Anthology* noted with pleasure his account of the aboriginal earthworks of the Ohio Valley, adding: "The subject is a very interesting one; and will no doubt, when thoroughly investigated, prove that the aboriginal inhabitants of America originated, like the rest of mankind, from Asia." *Monthly Anthology*, 1805, *2*: 595.

[36] Constantin F. Chasseboeuf, Comte de Volney, *A View of the Soil and Climate of the United States of America. . .* , Charles Brockden Brown, tr. (Philadelphia, 1804), 425, translator's note. Brown adds: "But though Volney will be thought by many to overrate the importance of his subject, enlightened minds must acknowledge that it is a curious and instructive one."

be confidently predicted, wrote Benjamin Silliman, "that until intelligent chemists are trained up at home, and induced to attempt the introduction and extension of the CHEMICAL ARTS, the United States will never attain to the pinnacle of national superiority, which Great Britain and France owe more to the *successful cultivation and application of natural science*, than the one does to the prowess of her armies, or the other to the triumphs of her marine." [37] Dr. Thomas Ewell elaborated the same theme somewhat more poetically:

> The time is now come, when the citizens of America should act entirely for themselves; when they should forever cease to depend on the caprice of foreigners, for the innumerable chemical compounds. This fair portion of the civilized world . . . must in all its parts be adorned with native chemists. The dignity of independence and the glory of usefulness, should rouse the love of science from the lethargic dispositions in America.[38]

Appeals to the utilitarian spirit of the age were doubtless effective, especially when linked to patriotism, but the prevailing emphasis on immediate practical utility could be a hindrance, as well as a help, to science. Despite the repeated claim that science was useful, the suspicion persisted in the public mind that scientists were more interested in abstract research than in practical application. The *Medical Repository* noted Henry M. Brackenridge's apprehension "lest the character of a man of science should be fastened upon him." The editors regretted his attitude but conceded that, "in a state of society like ours, where a mere suspicion of possessing eminent attainments of this kind, too often lessens the confidence reposed in an individual as a man of business, it is certainly discreet to avoid the imputation." [39] De Witt Clinton, addressing the Literary and Philosophical Society of New York in 1814, deprecated American preoccupation with the pursuit of gain, finding in it one explanation of the fact that "the enterprising spirit which distinguishes our national character has exhibited itself in every shape except that of a marked devotion to science." [40]

Scientists were careful, therefore, to stress the moral and religious value of science as well as its practical utility. In an age when mere entertainment was suspect, science could be recommended as the highest form of "rational amusement." To a public which believed firmly that nature was the creation of an omnipotent, omniscient and benevolent God, the study of God's works could be represented as a duty, not merely a pleasure. Natural theology was in the air. Smellie's *Philosophy of Natural History* and Paley's *Natural Theology* were available in American editions at the booksellers, along with translations of French and German works of the same kind. Although Americans wrote few works of this kind themselves, the teleological view of nature permeated all their writing, whether scientific or didactic, popular or learned, Christian or deist.[41] To be sure, the question of the plenary inspiration of Scripture was highly controversial, and Peale found that he had to soft-pedal his deism in order to avoid offending museum patrons. In the book business, however, controversy was the life of trade. Did geology disprove the

[37] Benjamin Silliman, "Notes . . . to Henry's Chemistry," in Henry, *Epitome*, ii.

[38] Thomas Ewell, *Discourses*, 30–31.

[39] Review of H. M. Brackenridge's *Views of Louisiana* (Pittsburgh, 1814) in *The Medical Repository*, 1815, *2*: 259–260.

[40] De Witt Clinton, *An Introductory Discourse, Delivered Before the Literary and Philosophical Society of New York, on the Fourth of May, 1814* (New York, 1815), 4.

[41] See, for example, the "Preface" to Noah Webster's *History of Animals* (New Haven, 1812): "The descriptions here given are interspersed with some amusing anecdotes; and with occasional moral and religious reflections, calculated to lead the young mind to contemplate the character, and to admire the wisdom, power and goodness of the divine author of all life, and of all the order, beauty and harmony that are visible among created beings."

Deluge? Could the Negro and the American Indian be derived from Adam and Eve? Had Linnaean biology discredited the idea of spontaneous generation? Questions of this kind received great attention. Samuel Miller's *Brief Retrospect of the Eighteenth Century* devoted nearly as much space to showing the harmony of science and religion as to describing scientific progress. "In Massachusetts," reported the *Monthly Anthology*, "the *Retrospect* has become a fashionable book . . . . Religious people, especially, will be fond of reading a book which makes every event and every work illustrative of the belief of the scriptures . . . ." [42]

Intellectual curiosity and love of the marvellous also contributed to popular interest in science. Generally speaking, however, American scientists made little use of the latter appeal, regarding it as meretricious and unworthy. They addressed themselves to the patriotic citizen and the serious student. Charles Willson Peale was one of the few who realized that the common man's astonishment at curious and unusual things might be turned to the advantage of science if exploited with restraint. Through it the public might be led toward a more solid and enduring interest in nature at the same time that science benefited from the growth of a scientific museum. But Peale never succeeded in convincing the scientists of Philadelphia, much less the general public, that his idea was a feasible one, worthy of governmental support. It may well be that his failure to win his battle for public support was due as much to the distrust of American scientists for popular science as to lack of interest on the part of the public.

Thus, the initial question recurs: what was the extent of popular interest in science in the early national period? To answer this question satisfactorily one would need a more precise definition of popular interest, a more reliable index to its intensity, a much fuller examination of the bearing of science on other fields of endeavor, and a closer study of the intellectual interests of particular groups of Americans.[43] The most that can be said on the basis of the evidence herein presented is that the scientists of Jefferson's day found their countrymen all too little interested in science, and that they labored as best they could to cultivate a taste for it among the educated classes by appeals to patriotism and civic pride, to natural theology and the utilitarian spirit of the age. A few enthusiasts, notably Charles Willson Peale and John Griscom, attempted to reach beyond the charmed circle of gentility to the artisan and the farmer. Their efforts, though only partially successful, heralded the coming of a new and more democratic age in American science.

[42] *Monthly Anthology*, 1804, *1*: 364–365. For a case study of the stimulus given to anthropological researches by an interest in their bearing on the doctrine of the plenary inspiration of Scripture, see the present author's "American Debate on the Negro's Place in Nature, 1780–1815," *Journal of the History of Ideas*, 1954, *15*: 384–396.

[43] See, for example, Harry Hayden Clark's "The Influence of Science on American Ideas, from 1775 to 1809," *Trans. Wisc. Acad. Sci., Arts, Letters*, 1944, *35*: 305–349.

Library of Congress Cataloging in Publication Data

Main entry under title:

Early American science.

(History of science, selections from Isis)
Selected articles from
Isis.
Includes bibliographies.
1. Science--History--United States--Addresses, essays, lectures. 2. Scientists--United States--Addresses, essays, lectures. I. Hindle, Brooke. II. History of science. III. Isis.
Q127.U6E17 509'.73 76-3667
ISBN 0-88202-151-6

ISBN 0-88202-150-8 pbk.